ATLAS
Grade 7 Math Practice

GET DIGITAL ACCESS TO

 2 ATLAS Practice Tests

 Personalized Study Plans

REGISTER NOW

Link	QR Code

Visit the link below for online registration

lumoslearning.com/a/tedbooks

Access Code: ATLASG7M-47613-P

Arkansas Teaching and Learning Assessment System Test Prep: 7th Grade Math Practice Workbook and Full-length Online Assessments: ATLAS Study Guide

Contributing Editor - Aaron Spencer
Contributing Editor - Nikki McGee
Executive Producer - Mukunda Krishnaswamy
Program Director - Anirudh Agarwal
Designer and Illustrator - Sowmya R.

ISBN 13: 978-1959697206

Printed in the United States of America

FOR SCHOOL EDITION AND PERMISSIONS, CONTACT US

LUMOS INFORMATION SERVICES, LLC

 PO Box 1575, Piscataway, NJ 08855-1575
 www.LumosLearning.com

 Email: support@lumoslearning.com
 Tel: (732) 384-0146
Fax: (866) 283-6471

Lumos Learning
Step Up Your Skills

INTRODUCTION

This book is specifically designed to improve student achievement on the Arkansas Teaching and Learning Assessment System(ATLAS). Students perform at their best on standardized tests when they feel comfortable with the test content as well as the test format. Lumos online practice tests are meticulously designed to mirror the state assessment. They adhere to the guidelines provided by the state for the number of sessions and questions, standards, difficulty level, question types, test duration and more.

Based on our decade of experience developing practice resources for standardized tests, we've created a dynamic system, the Lumos Smart Test Prep Methodology. It provides students with realistic assessment rehearsal and an efficient pathway to overcoming each proficiency gap.

Use the Lumos Smart Test Prep Methodology to achieve a high score on the ATLAS.

Lumos Smart Test Prep Methodology

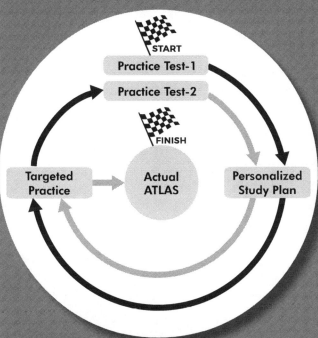

1 The student takes the first online diagnostic test, which assesses proficiency levels in various standards.

2 StepUp generates a personalized online study plan based on the student's performance.

3 The student completes targeted practice in the printed workbook and marks it as complete in the online study plan.

4 The student then attempts the second online practice test.

5 StepUp generates a second individualized online study plan.

6 The student completes the targeted practice and is ready for the actual ATLAS.

Table of Contents

Chapter 1
Number Concepts & Computations

Lesson 1: Add and Subtract Rational Numbers

1. If p + q has a value that is exactly $\frac{1}{3}$ less than p, what is the value of q?
   - (A) $\frac{-1}{3}$
   - (B) $\frac{2}{5}$
   - (C) $\frac{1}{3}$
   - (D) $\frac{-2}{5}$

2. What is the sum of k and the opposite of k?
   - (A) 2k
   - (B) k + 1
   - (C) 0
   - (D) –1

3. If p + q has a value of $\frac{12}{5}$, and p has a value of $\frac{4}{5}$, what is the value of q?
   - (A) $\frac{5}{8}$
   - (B) $\frac{8}{5}$
   - (C) $\frac{1}{3}$
   - (D) $\frac{3}{2}$

4. **What is the value of b in the diagram?**

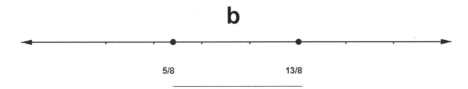

b

5/8   13/8

Ⓐ 0.8
Ⓑ 1
Ⓒ 1.8
Ⓓ 2.2

5. t has a value of $\dfrac{5}{2}$. p is the sum of t and v, and p has a value of 0. What is the value of v?

Ⓐ $\dfrac{-1}{3}$

Ⓑ 4

Ⓒ 2.5

Ⓓ $\dfrac{-5}{2}$

6. **What is the value of n in the diagram?**

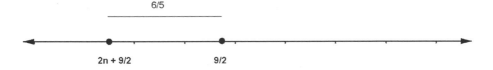

6/5

2n + 9/2   9/2

Ⓐ $\dfrac{6}{5}$

Ⓑ 0.6

Ⓒ $\dfrac{-3}{5}$

Ⓓ 2.1

7. **What is the sum of 10 and –10?**

Ⓐ 20
Ⓑ 0
Ⓒ 1
Ⓓ –20

8. If the value of r is $\dfrac{-1}{3}$ and t has a value of $\dfrac{-1}{2}$, will t + r be to the right or left of t on a number line, and why?

    Ⓐ To the left because the absolute value of r is less than the absolute value of t.
    Ⓑ To the right because the absolute value of r is less than the absolute value of t, and both numbers are negative.
    Ⓒ To the left because r is a negative value being added to t.
    Ⓓ To the left because the sum of two negative numbers is always negative.

9. If the value of A is 3, and the value of B is $\dfrac{-2}{3}$, how far apart will A and B be on a number line?

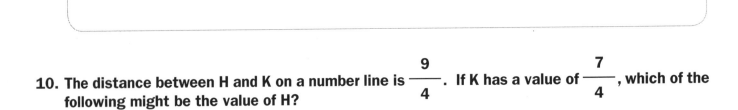

10. The distance between H and K on a number line is $\dfrac{9}{4}$. If K has a value of $\dfrac{7}{4}$, which of the following might be the value of H?

    Ⓐ $\dfrac{2}{4}$

    Ⓑ $\dfrac{9}{4}$

    Ⓒ - 4

    Ⓓ $\dfrac{-1}{2}$

11. Solve $\dfrac{9}{14} - \dfrac{3}{14}$ and indicate it by shading the relevant boxes.

12. Enter the missing part of the addition or subtraction problem into the blanks in the table. Express fractions as mixed fractions wherever applicable.

$-\dfrac{11}{3}$	$+$	$\dfrac{2}{13}$	$-\dfrac{9}{13}$
3.2		-5.7	8.9
$-2\dfrac{3}{4}$	$-$		4
	$+$	8.1	-5.1

CHAPTER 1 → Lesson 2: Rational Numbers, Addition & Subtraction

1. **Evaluate: 25 + 2.005 - 7.253 - 2.977**

    (A) -16.775
    (B) 16.775
    (C) 167.75
    (D) 1.6775

2. **Add and/or subtract as indicated :** $-3\dfrac{4}{5} + 9\dfrac{7}{10} - 2\dfrac{11}{20} =$

    (A) $3\dfrac{7}{20}$

    (B) $4\dfrac{7}{10}$

    (C) $4\dfrac{9}{20}$

    (D) $3\dfrac{1}{20}$

3. **Linda and Carrie made a trip from their hometown to a city about 200 miles away to attend a friend's wedding. The following chart shows their distances, stops and times. What part of their total trip did they spend driving?**

3hr	driving
15 min	rest stop
1 1/2 hr	driving
1 hr	rest stop
20 min	driving

    (A) $\dfrac{4}{5}$

    (B) $\dfrac{2}{5}$

    (C) $\dfrac{58}{73}$

    (D) $\dfrac{99}{100}$

4. If $a = \dfrac{5}{6}$, $b = -\dfrac{2}{3}$ and $c = -1\dfrac{1}{3}$, find a - b - c.

     Ⓐ $-1\dfrac{1}{6}$

     Ⓑ $2\dfrac{5}{6}$

     Ⓒ $-2\dfrac{1}{6}$

     Ⓓ $-2\dfrac{5}{6}$

5. If Ralph ate half of his candy bar followed by half of the remainder followed by half of that remainder, what part was left?

     Ⓐ $\dfrac{1}{4}$

     Ⓑ $\dfrac{1}{8}$

     Ⓒ $\dfrac{1}{6}$

     Ⓓ $\dfrac{1}{3}$

6. Mary is making a birthday cake to surprise her mom. She needs $3\dfrac{1}{2}$ cups of flour but she only has $\dfrac{1}{3}$ cup. How much more flour does she need?

7. Ricky purchased shoes for $159.95 and then exchanged them at a buy 1, get 1 half-off sale. The shoes that he purchased on his return trip were $74.99 and $68.55. How much did he receive back from the store after his second transaction? Note: The half-price applies to the cheaper of the two pairs.

    Ⓐ $37.50
    Ⓑ $68.55
    Ⓒ $34.28
    Ⓓ $50.68

8. Simplify the following expression:

    $3.24 - 1.914 - 6.025 + 9.86 - 2.2 + 5\frac{1}{2}$ =

    Ⓐ -8.461
    Ⓑ 8.461
    Ⓒ -11.259
    Ⓓ 11.259

9. John had $76.00. He gave Jim $42.45 and gave Todd $21.34. John will receive $14.50 later in the evening. How much money will John have later that night?

    Ⓐ $25.71
    Ⓑ $26.67
    Ⓒ $26.71
    Ⓓ $24.71

10. Jeri has had a savings account since she entered first grade. Each month of the first year she saved $1.00. Each month of the second year she saved $2.00. This pattern of adding one additional dollar each month for an entire year continued until she was saving $10.00 per month in the tenth year. How much does she have saved at the end of ten years?

    Ⓐ $660
    Ⓑ $648
    Ⓒ $636
    Ⓓ $624

11. Solve : $\frac{3}{7} + \left(-\frac{5}{7}\right)$ Write your answer in the box given below.

12. Which expressions equal - $\frac{3}{4}$? Select all the correct answers.

Ⓐ $\frac{1}{8} - \frac{7}{8}$

Ⓑ $\frac{7}{8} - \frac{1}{8}$

Ⓒ $-\frac{6}{4} + \frac{3}{4}$

Ⓓ $\frac{1}{4} + \frac{1}{8}$

13. Read the number sentences below and match it with the correct associated property.

	Associative Property of Addition	Inverse Property of Addition	Identity Property of Addition
$\frac{2}{5} + 0 = \frac{2}{5}$	○	○	○
$\frac{1}{4} + \left(\frac{2}{3} + \frac{7}{8}\right) = \left(\frac{1}{4} + \frac{2}{3}\right) + \frac{7}{8}$	○	○	○
$\frac{6}{7} + \left(-\frac{6}{7}\right) = 0$	○	○	○

CHAPTER 1 → Lesson 3: Additive Inverse and Distance Between Two Points on a Number Line

1. **Which of the following is the same as 7 – (3 + 4)?**

    Ⓐ 7 + (–3) + (–4)
    Ⓑ 7 +(–3 + 4)
    Ⓒ 7 + 7
    Ⓓ –7 – 7

2. **Which of the following expressions represents the distance between the two points?**

    Ⓐ | 4 - 3 |
    Ⓑ (-3) -4
    Ⓒ | (-3) - 4 |
    Ⓓ 4 - 3

3. **Kyle and Mark started at the same location. Kyle traveled 5 miles due east, while Mark traveled 3 miles due West. How far apart are they?**

    Ⓐ 2 miles
    Ⓑ 8 miles
    Ⓒ 15 miles
    Ⓓ 12 miles

4. **The distance between G and H on the number line is | 5 - (- 2) |. What might be the coordinate of H?**

    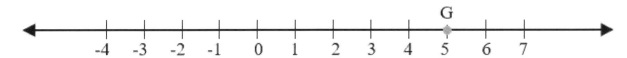

    Ⓐ 2
    Ⓑ 7
    Ⓒ 6
    Ⓓ –2

5. Which of the following is the same as 2x – 3y – z?

    Ⓐ 2x – (3y – z)
    Ⓑ 2x + (–3y) – z
    Ⓒ 2x + 3y – z
    Ⓓ (–2x) – 3y – z

6. If Brad lives 5 blocks north of the park and Easton lives 7 blocks south of the park, which of the following correctly represents an expression for the distance between their homes?

    Ⓐ | 5 - (-7) |
    Ⓑ 7 - 5
    Ⓒ (- 7) - 5
    Ⓓ | 5 + (- 7) |

7. For what numbers will the statement be true?: t – (w) = t + (–w)

    Ⓐ All positive numbers only
    Ⓑ All positive integers only
    Ⓒ All positive and negative integers, but not 0
    Ⓓ All real numbers

8. Which of the following is the same as 4 + (x – 3)?

    Ⓐ 4 + (–x) – 3
    Ⓑ (–4) + x + (–3)
    Ⓒ 4 + x + (–3)
    Ⓓ 1 + (–x)

9. Which of the following expressions represents the distance between the two points?

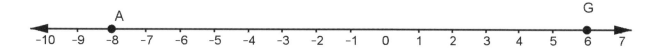

    Ⓐ |6-8|
    Ⓑ (- 8)+6
    Ⓒ (- 8)- 6
    Ⓓ |6 - (- 8)|

10. Which words best complete the statement?

    The distance between two numbers on a number line is the same as the _____ of their
    _____.

    Ⓐ Absolute value; difference
    Ⓑ Sum; squares
    Ⓒ Difference; squares
    Ⓓ Absolute value; sum

11. Use the number line to determine the distance between $-1\frac{1}{6}$ and $\frac{5}{6}$. Write your answer in
    simplest form in the box given below.

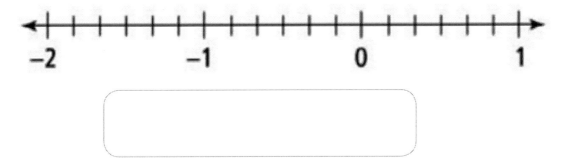

12. Which of these situations can be represented by the opposite of 34?

    Instruction: More than one option may be correct.

    Ⓐ The airplane ascends 34 feet.
    Ⓑ The elevator descends 34 feet.
    Ⓒ The cost of the coat is $34 more than expected.
    Ⓓ Jack removes 34 books from the library shelf.

13. State whether a number and its inverse are added in these expressions. If yes, select additive inverse is added. If no, select additive inverse is not added.

	Additive Inverse is added	Additive Inverse is not added
3 + (-3)	○	○
-2.2 + 2.2	○	○
1 + 1	○	○
$-\frac{4}{7} + \left(-\frac{4}{7}\right)$	○	○

CHAPTER 1 → Lesson 4: Converting Between Rational Numbers and Decimals

1. **Convert to a decimal:** $\dfrac{7}{8}$

    Ⓐ 0.78
    Ⓑ 0.81
    Ⓒ 0.875
    Ⓓ 0.925

2. **Convert to a decimal:** $\dfrac{5}{6}$

    Ⓐ 0.8333333...
    Ⓑ 0.56
    Ⓒ 0.94
    Ⓓ 0.8

3. **How can you tell that the following number is a rational number?**

    0.251

    Ⓐ It is a rational number because the decimal terminates.
    Ⓑ It is a rational number because there is a value of 0 in the ones place.
    Ⓒ It is a rational number because the sum of the digits is less than 10.
    Ⓓ It is a rational number because it is not a repeating decimal.

4. **A group of 11 friends ordered 4 pizzas to share. They divided the pizzas up evenly and all ate the same amount. Express in decimal form the portion of a pizza that each friend ate.**

    Ⓐ 0.36363636...
    Ⓑ 0.411
    Ⓒ 0.14141414...
    Ⓓ 0.48

5. **How can you tell that the following number is a rational number?**

    0.133333...

    Ⓐ It is rational because the decimal does not terminate.
    Ⓑ It is rational because the decimal repeats over and over.
    Ⓒ It is rational because the number is a factor of 1.
    Ⓓ It is NOT rational because the decimal does not terminate.

6. Convert to a decimal: $2\dfrac{2}{9}$

   Ⓐ 2.92299229...
   Ⓑ 2.2222...
   Ⓒ 2.35
   Ⓓ 2.4835

7. How can you tell that the following number is not a rational number?:

   2.4876352586582142597868...

   Ⓐ It is not rational because the same digit never occurs twice in a row in the decimal.
   Ⓑ It is not rational because the decimal does not terminate or repeat.
   Ⓒ It is not rational because it is greater than 1.
   Ⓓ It is not rational because it is not a factor of 5.

8. Convert to a decimal: $\dfrac{11}{25}$

   Ⓐ 0.4444444...
   Ⓑ 0.472
   Ⓒ 0.369
   Ⓓ 0.44

9. The track at Haley's school is a third of a mile in length. She ran 14 laps on it after school one day. Express the number of miles she ran in decimal form.

   Ⓐ 4.56
   Ⓑ 3.85
   Ⓒ 4.6666...
   Ⓓ 4.725

10. Convert to a decimal: $\dfrac{34}{99}$

   Ⓐ 0.943
   Ⓑ 0.343434...
   Ⓒ 0.394
   Ⓓ 0.439439...

11. In the library, there are 230 fiction books, 120 nonfiction books, and 30 magazines. Write the ratio of magazines to nonfiction books as a decimal in the box given below.

12. **Which of the below statements are true? Select all the correct answer choices.**

Ⓐ $\frac{1}{4}$, 0.25, and 25% are equivalent

Ⓑ $\frac{1}{5}$, 0.20, and 20% are equivalent

Ⓒ $\frac{1}{4}$, 0.14, and 14% are equivalent

Ⓓ $\frac{1}{5}$, 0.20, and 2% are equivalent

Ⓔ $\frac{1}{4}$, 0.25, and 2.5% are equivalent

13. **Convert the rational numbers to decimals. Mark whether the decimals are repeating decimals or terminating decimals.**

	Repeating Decimal	Terminating Decimal
$\frac{3}{4}$	◯	◯
$\frac{2}{3}$	◯	◯
$\frac{1}{9}$	◯	◯
$\frac{2}{8}$	◯	◯

CHAPTER 1 → Lesson 5: Solving Real World Problems

1. Andrew has $9.39 but needs $15.00 to make a purchase. How much more does he need?

   Ⓐ $6.39
   Ⓑ $5.61
   Ⓒ $5.39
   Ⓓ $6.61

2. Ben has to unload a truck filled with 25 bags of grain for his horses. Each bag weighs 50.75 pounds.

   How many total pounds does he have to move?

   Ⓐ 12,687.50 pounds
   Ⓑ 1,268.75 pounds
   Ⓒ 126.875 pounds
   Ⓓ 1250 pounds

3. A Chinese restaurant purchased 1528.80 pounds of rice. If they received 50 identical bags, how much rice was in each bag?

   Ⓐ 30.576 pounds
   Ⓑ 305.76 pounds
   Ⓒ 3.0576 pounds
   Ⓓ None of the above.

4. Leila stopped at the coffee shop on her way to work. She ordered 2 bagels, 3 yogurts, and 1 orange juice. Bagels were $0.69 each, yogurts were $1.49 each, and orange juice was $1.75. What was Leila's total bill?

   Ⓐ $7.60
   Ⓑ $3.93
   Ⓒ $5.16
   Ⓓ $5.42

5. Mickey bought pizza and sodas for himself and four of his friends. The pizza was $17.49, and 5 sodas were $1.19 each.

   If the pizza is sliced into 10 equal slices and each person eats 2 slices and drinks one soda, what is the cost to each person?

   Ⓐ $2.94
   Ⓑ $4.13
   Ⓒ $3.50
   Ⓓ $4.69

6. Alan has to keep within a $15.00 budget. Tax is 6.5%. What is Alan's total if he buys 1 notebook, 1 pack of paper, 1 set of dividers, 2 pens and 5 pencils?

3-Ring Notebook	$5.69
Notebook Paper	$1.39
Dividers	$1.45
Pens	$1.19
Pencils	$0.50

Ⓐ $10.88
Ⓑ $14.28
Ⓒ $13.41
Ⓓ $15.00

7. Sammy is mowing the lawn. The lawn is 30 ft by 20 ft. Sammy cut a strip 5 ft by 10 ft and ran out of gas. How much more does he need to mow?

Ⓐ 250 sq ft
Ⓑ 550 sq ft
Ⓒ 50 sq ft
Ⓓ 600 sq ft

8. Dustin has leased 5 acres of land to raise produce for the farmers' market. He has already planted $\frac{5}{8}$ of the land.

How many more acres does he need to plant?

Ⓐ $3\frac{1}{8}$ acres
Ⓑ 4 acres
Ⓒ $1\frac{7}{8}$ acres
Ⓓ $\frac{3}{8}$ acres

9. Taylor bought a bag of marbles weighing 5.25 lb. Before he got to the car, the bag broke, spilling many marbles. To find out if he had recovered all of his marbles, he weighed the bag at home.

He found that his bag of marbles now weighed 4.98 lb. What was the weight of the lost marbles?

Ⓐ 0.27 lb
Ⓑ 2.7 lb
Ⓒ 0.173 lb
Ⓓ 1.73 lb

10. Which of the following is a correct statement about the multiplication of two integers?

Ⓐ If the signs are both negative, multiply the numbers and give the answer a negative sign.
Ⓑ If the signs are both negative, multiply the numbers and give the answer a positive sign.
Ⓒ If the signs are unlike, multiply the numbers and give the answer a positive sign.
Ⓓ If the signs are unlike, multiply the numbers and give the answer the sign of the larger absolute value.

11. Robert has $\frac{4}{7}$ of a bucket of water. John has $\frac{2}{3}$ of a bucket of water. If Robert and John combine their buckets of water, how many buckets of water will they have? Write your answer in simplest form. Write your answer in the box given below.

12. Frank lives $\frac{1}{2}$ blocks east of Mary. Mary lives $\frac{3}{5}$ blocks east of school. How many blocks east of school does Frank live? There can be more than 1 correct answer. Select all the correct ones.

Ⓐ $1\frac{1}{10}$

Ⓑ $\frac{11}{10}$

Ⓒ $\frac{3}{5}$

Ⓓ $\frac{4}{2}$

Ⓔ $\frac{4}{7}$

13. Solve each equation and mark if the answer is negative, positive or zero.

	Negative	Zero	Positive
$-\frac{6}{7} - \left(-\frac{6}{7}\right)$	○	○	○
$-5 - \frac{3}{5}$	○	○	○
$\frac{3}{4} - \frac{1}{5}$	○	○	○

CHAPTER 1 → Lesson 6: Strategies for Adding and Subtracting Rational Numbers

1. **What property is illustrated in the equation?:**

    $7 + \dfrac{1}{2} = \dfrac{1}{2} + 7$

    - Ⓐ Commutative property of addition
    - Ⓑ Associate property of addition
    - Ⓒ Distributive property
    - Ⓓ Identity property of addition

2. **Which is a valid use of properties to make the expression easier to calculate?**

    $92 - 8 = ?$

    - Ⓐ $90 - 2 - 8$
    - Ⓑ $82 - (10 - 8)$
    - Ⓒ $82 + (10 - 8)$
    - Ⓓ $(80 + 2) - 8$

3. **Find the sum of the mixed numbers.**

    $4\dfrac{7}{8} + 7\dfrac{5}{8}$

    - Ⓐ $11\dfrac{5}{8}$
    - Ⓑ $12\dfrac{1}{2}$
    - Ⓒ $12\dfrac{7}{8}$
    - Ⓓ $11\dfrac{1}{2}$

4. **Which is a correct use of the distributive property?**

    - Ⓐ $3(4+2) = 3(4)+2$
    - Ⓑ $3(4+2) = (3 \times 4) \times (3 \times 2)$
    - Ⓒ $3(4+2) = 3(4)+3(2)$
    - Ⓓ All of the above

5. **Name the property illustrated in the equation:**

    (5 + 8) + 2 = 5 + (8 + 2)

    Ⓐ Distributive property
    Ⓑ Associative property of addition
    Ⓒ Commutative property of addition
    Ⓓ Identity property of addition

6. **Which is a valid use of properties to make the expression easier to calculate?**

    7 – 2.45

    Ⓐ (6 – 2) + (1 – 0.45)
    Ⓑ (7 – 2) + 0.45
    Ⓒ 5 + 0.45
    Ⓓ 7 – (0.45 – 2)

7. **Write the improper fraction as a mixed number.**

    $$\frac{31}{3}$$

    Ⓐ $3\frac{1}{3}$

    Ⓑ $30\frac{1}{3}$

    Ⓒ $31\frac{1}{3}$

    Ⓓ $10\frac{1}{3}$

8. **Find the difference between $16\frac{3}{4}$ - $9\frac{7}{8}$**

    Ⓐ $7\frac{1}{8}$

    Ⓑ $6\frac{7}{8}$

    Ⓒ $7\frac{7}{8}$

    Ⓓ $6\frac{1}{8}$

9. **What property is illustrated in the equation?:**

   (5 + 3) + 0 = 5 + 3

   - Ⓐ Identity property of addition
   - Ⓑ Associative property of addition
   - Ⓒ Commutative property of addition
   - Ⓓ Distributive property

10. **Which is a valid use of properties to make the expression easier to calculate? Circle the correct answer choice.**

$$\frac{20}{7} - 1\frac{3}{7} =$$

   - Ⓐ $\frac{20}{7} - 1 + \frac{3}{7}$
   - Ⓑ $\frac{13}{7} - 1 + \frac{3}{7}$
   - Ⓒ $\left(\frac{13}{7} - \frac{3}{7}\right) + (1 - 1)$
   - Ⓓ $\frac{20}{7} - \frac{1}{7} - \frac{3}{7}$

11. **Which of the pairs of numbers are 13.7 units apart on a number line? More than 1 answer may be correct. Mark all the correct answers.**

   - Ⓐ - 26.3 and - 12.6
   - Ⓑ - 26.3 and 12.6
   - Ⓒ 26.3 and - 12.6
   - Ⓓ - 3.2 and 10.5

12. **Write the correct absolute value expression or distance description into the blanks in the table.**

Distance between 4 and -1	|-1-4|	|4-(-1)|
Distance between -4 and -1		|-4-(-1)|
Distance between 4 and 1	|1-4|	
	|1-(-4)|	|-4-1|

CHAPTER 1 → Lesson 7: Strategies for Multiplying and Dividing Rational Numbers

1. **Which of these multiplication expressions is equivalent to the division expression below?**

$$-\frac{4}{23} \div \frac{7}{58}$$

Ⓐ $-\frac{23}{4} \times \frac{58}{7}$

Ⓑ $-\frac{4}{23} \times \frac{58}{7}$

Ⓒ $-\frac{23}{4} \times \frac{7}{58}$

Ⓓ $-\frac{4}{23} \times \frac{7}{58}$

2. **Which property is illustrated in the following statement?**
   (5)(4)(7) = (7)(5)(4)

   Ⓐ Triple multiplication property
   Ⓑ Distributive property
   Ⓒ Commutative property of multiplication
   Ⓓ Associative property of multiplication

3. **Find the quotient: $4 \div \dfrac{1}{2}$**

   Ⓐ 4
   Ⓑ 8
   Ⓒ 2
   Ⓓ 16

4. **Which of the following is a helpful and valid way to rewrite the expression for evaluation :**
   3(123)

   Ⓐ 3(100) + 3(20) + 3(3)
   Ⓑ 3(100) + 23
   Ⓒ 100 + 3(23)
   Ⓓ 300 + 3(20) + 3

5. Which multiplication expression are equivalent to $\dfrac{7}{8} \div \dfrac{1}{15}$?

    Ⓐ $\dfrac{7}{8}$ x 15

    Ⓑ $\dfrac{7}{8}$ x $\dfrac{1}{15}$

    Ⓒ $\dfrac{8}{7}$ ÷ $\dfrac{1}{15}$

    Ⓓ $\dfrac{8}{7}$ x $\dfrac{1}{15}$

6. Which property is illustrated in the following statement?

    $(2 \cdot 3) \cdot 6 = 2 \cdot (3 \cdot 6)$

    Ⓐ Associate property of multiplication
    Ⓑ Commutative property of multiplication
    Ⓒ Identity property of multiplication
    Ⓓ Reflexive property

7. Find the product: $\dfrac{3}{2} \ast \dfrac{7}{6}$

    Ⓐ $\dfrac{10}{12}$

    Ⓑ $\dfrac{21}{6}$

    Ⓒ $\dfrac{7}{4}$

    Ⓓ $\dfrac{10}{8}$

8. **Find the quotient:** $\dfrac{6}{5} \div \dfrac{3}{25}$

  Ⓐ $\dfrac{2}{5}$

  Ⓑ 2

  Ⓒ $\dfrac{5}{2}$

  Ⓓ 10

9. **Find the product:** $\left(2\dfrac{1}{3} \right)\left(3\dfrac{1}{2} \right)$

  Ⓐ $\dfrac{49}{6}$

  Ⓑ $6\dfrac{1}{6}$

  Ⓒ $\dfrac{6}{5}$

  Ⓓ $5\dfrac{2}{5}$

10. **Find the quotient:** $\left(5\dfrac{1}{4} \right) \div \left(1\dfrac{1}{2} \right)$

  Ⓐ $\dfrac{5}{8}$

  Ⓑ $\dfrac{7}{2}$

  Ⓒ 45

  Ⓓ 7

11. Fill in the table with the correct reciprocals.

$\dfrac{6}{4}$	$\dfrac{4}{6}$
	$\dfrac{-19}{6}$
$\dfrac{5}{-64}$	
	4

CHAPTER 1 → Lesson 8: Rational Numbers as Quotients of Integers

1. **Which of the following division problems CANNOT be completed?**

   Ⓐ $10 \div 0$
   Ⓑ $155 \div (-3)$
   Ⓒ $\dfrac{2}{3} + \dfrac{1}{4}$
   Ⓓ $0 \div 5$

2. **Which of the following is NOT equivalent to the given value?:**

   $-\dfrac{2}{3}$

   Ⓐ $\dfrac{-2}{(3)}$

   Ⓑ $\dfrac{-2}{(-3)}$

   Ⓒ $\dfrac{2}{(-3)}$

   Ⓓ All are equivalent values

3. **Jared hiked a trail that is 12 miles long. He hiked the trail in section that were 1.5 miles each. In how many sections did he complete the hike?**

   Ⓐ 12
   Ⓑ 10
   Ⓒ 8
   Ⓓ 6

4. **Greg is able to run a mile in 8 minutes. He ran at that pace for t minutes. What does the following expression represent?**

   $\dfrac{t}{8}$

   Ⓐ The number of miles that Greg ran.
   Ⓑ The number of hours that Greg ran.
   Ⓒ The average pace at which Greg ran.
   Ⓓ The amount of time it took Greg to run 8 miles.

5.   What value could x NOT be in the following expression?:

$$\frac{5+x}{x}$$

    Ⓐ 5
    Ⓑ 1
    Ⓒ –5
    Ⓓ 0

6.   Rose is filling her swimming pool with water. She needs to pump 2000 gallons into the pool, and the water flows at a rate of r gallons per hour. Which of the following expresses the amount of time it will take to fill the pool?

    Ⓐ $\dfrac{2000}{r}$

    Ⓑ 2000*r*

    Ⓒ 2000 + *r*

    Ⓓ $\dfrac{r}{2000}$

7.   If t = –2, and v = –4, which of the following is equal to $\dfrac{t}{v}$?

    Ⓐ $\dfrac{t}{4}$

    Ⓑ $\dfrac{4}{2v}$

    Ⓒ $\dfrac{-t}{4}$

    Ⓓ $\dfrac{-6}{-3v}$

8. If, $\dfrac{3a}{b} = 12$, what is the value of $\left(-\dfrac{a}{b}\right)$?

Ⓐ   - 4

Ⓑ $\dfrac{3}{4}$

Ⓒ $-\dfrac{3}{4}$

Ⓓ $\dfrac{1}{4}$

9. **Which terms best complete the statement?**

   A number is rational if it can be expressed as a _____ of _____.

   Ⓐ   sum; integers
   Ⓑ   quotient; integers
   Ⓒ   difference; prime numbers
   Ⓓ   product; variables

10. **Kathy is laying stepping stones in her garden. The stones are 8 inches long, and she wants to create a path that is 10 feet long. How many stones will she need?**

   Ⓐ   10 stones
   Ⓑ   80 stones
   Ⓒ   15 stones
   Ⓓ   1.25 stones

11. **Complete the fraction:** $\dfrac{?}{3} = -5$. **Write your answer in the box given below.**

12. Which of the quotients are equivalent to $-\dfrac{6}{7}$?

    Note that more than one option may be correct. Select all the correct options.

    Ⓐ $\dfrac{6}{7}$

    Ⓑ $-\dfrac{-6}{7}$

    Ⓒ $-\dfrac{-6}{-7}$

    Ⓓ $\dfrac{-6}{-7}$

    Ⓔ $\dfrac{6}{-7}$

    Ⓕ $-\dfrac{6}{7}$

13. Mark the boxes to indicate whether the quotient is positive, negative, zero, or undefined.

	Positive	Negative	Zero	Undefined
$-50 \div 5$	○	○	○	○
$\dfrac{0}{-4}$	○	○	○	○
$\dfrac{-35}{-7}$	○	○	○	○
$\dfrac{47}{0}$	○	○	○	○

End of Number Concepts & Computations

Chapter 1
Number Concepts & Computations

Lesson 1: Add and Subtract Rational Numbers

Question No.	Answer	Detailed Explanations
1	A	The fact that p + q is less than p tells us that q is negative. p + q is exactly $\frac{1}{3}$ less than p means, p + q = p - $\frac{1}{3}$. Therefore, q is -$\frac{1}{3}$.
2	C	Anytime that you add a number and its opposite, the sum is always 0.
3	B	The difference between $\frac{12}{5}$ and $\frac{4}{5}$ will be the value of q. The difference is $\frac{8}{5}$.
4	B	b is the length of the distance between $\frac{5}{8}$ and $\frac{13}{8}$. That difference is $\frac{8}{8}$, or 1.
5	D	Because t and v have a sum of 0, we know that they are opposite values. If t is $\frac{5}{2}$, then v is - $\frac{5}{2}$.
6	C	The distance between the two points is $\frac{6}{5}$. We also see that the difference in their given values is 2n. Because adding 2n takes us to the left of the other point, we know that 2n is a negative value of $\frac{6}{5}$. n must be half that value, or - $\frac{3}{5}$.
7	B	The sum of a number and its opposite is always 0.
8	C	If you add a negative number to t, you will always move left on the number line. So, t + r will be to the left of t because r is a negative number being added to t.

Question No.	Answer	Detailed Explanations
9	D	The distance between the two points, 3 and $-\frac{2}{3}$ is $3 - (-\frac{2}{3}) = 3 + \frac{2}{3} = \frac{11}{3}$
10	D	There are two possibilities of points that are $\frac{9}{4}$ away from $\frac{7}{4}$. One of the points is $\frac{9}{4}$ more than $\frac{7}{4}$, which would be $\frac{16}{4}$ (not an answer option). The other point is $\frac{9}{4}$ less than $\frac{7}{4}$, which is $-\frac{2}{4}$, or $-\frac{1}{2}$.
11	6	$\frac{9}{14} - \frac{3}{14} = \frac{6}{14}$ i.e. 6 out of 14. Therefore, shade 6 cells.

12

$-\frac{11}{3}$	+	$\frac{2}{13}$	$-\frac{9}{13}$
3.2	-	-5.7	8.9
$-2\frac{3}{4}$	-	$-6\frac{3}{4}$	4
- 13.2	+	8.1	-5.1

Use the rules of adding and subtracting positive and negative rational numbers to solve. Subtracting a negative can be changed to adding a positive number.

Lesson 2: Rational Numbers, Addition & Subtraction

Question No.	Answer	Detailed Explanations
1	B	Remember: adding and subtracting rational numbers works just like integers. If you need to carry or borrow, the rules remain the same. 25 + 2.005 - 7.253 - 2.977 27.005 - 7.253 - 2.977 , 19.752 - 2.977 = 16.775
2	A	$-3\frac{4}{5} + 9\frac{7}{10} - 2\frac{11}{20} = -3\frac{16}{20} + 9\frac{14}{20} - 2\frac{11}{20}$ $-5\frac{27}{20} + 9\frac{14}{20} = -5\frac{27}{20} + 8\frac{34}{20}$, $3\frac{7}{20}$ is the correct answer.
3	C	They spent 6 hours and 5 min total on their trip. Of that time, 4 hours and 50 minutes were spent driving. $4\frac{5}{6}$ hrs driving out of $6\frac{1}{12}$ hours. Converting into improper fractions we get $(\frac{29}{6})$ out of $(\frac{73}{12})$. $(\frac{29}{6}) \times (\frac{12}{73}) = \frac{58}{73}$ hours spent driving
4	B	If $a = \frac{5}{6}$, $b = -\frac{2}{3}$ and $c = -1\frac{1}{3}$, find a - b - c. $\frac{5}{6} - (-\frac{2}{3}) - (-1\frac{1}{3}) = \frac{5}{6} + \frac{4}{6} + 1\frac{2}{6} = 1\frac{11}{6} = 2\frac{5}{6}$. $2\frac{5}{6}$ is the correct answer.
5	B	Starting with a full candy bar, we take away 1/2 leaving 1/2. Then we take away 1/2 of the remaining half leaving 1/4 of the original candy bar. Then we take away half of the 1/4 leaving 1/8 of the original bar. 1/8 is the correct answer.
6		Number of more cup of flour required by Mary for making birthday cake = Number of cup of flour needed to prepare cake - Number of cup of flour Mary already has $= 3\frac{1}{2} - \frac{1}{3} = \frac{7}{2} - \frac{1}{3} = \frac{(21-2)}{6} = \frac{19}{6} = 3\frac{1}{6}$ Mary require $3\frac{1}{6}$ cup more flour to prepare the birthday cake.

Question No.	Answer	Detailed Explanations
7	D	On his second trip to the store he paid $74.99 plus half of $68.55. $74.99 + $34.28 = $109.27 $159.95 - 109.27 = $50.68 $50.68 is the correct answer.
8	B	First, we will change 5 1/2 to 5.5. Then we have 3.24 - 1.914 - 6.025 +9.86 -2.2 + 5.5 = 8.461 8.461 is the correct answer.
9	C	To solve this problem, list all the monetary values, along with the proper operation, before evaluating it. For this problem, words like "give" mean to subtract, while "receives" means to add. (1) 76.00 - 42.45 - 21.34 + 14.50 = (2) 33.55 - 21.34 + 14.50 = (3) 12.21 + 14.50 = (4) 26.71 Therefore, John will have $26.71 later that night.
10	A	Year 1 - $12.00, Year 2 - $24.00. Year 3 - $36.00. Year 4 - $48.00. Year 5 - $60.00. Year 6 - $72.00. Year 7 - $84.00. Year 8 - $96.00. Year 9 - $108.00. Year 10 - $120.00 Adding the totals for each year, we get $660.00. $660.00 is the correct answer.
11	$-\frac{2}{7}$	Since the denominators are the same just add the numerators. 3 + -5 = -2
12	A and C	$-\frac{6}{4} + \frac{3}{4}$ add the numerators and you get -3. So the fraction is $-\frac{3}{4}$ $\frac{1}{8} - \frac{7}{8}$ = 1 - 7 = -6 and $\frac{-6}{8}$ simplifies to $\frac{-3}{4}$

Question No.	Answer	Detailed Explanations

13			Associative Property of Addition	Inverse Property of Addition	Identity Property of Addition
		$\dfrac{2}{5} + 0 = \dfrac{2}{5}$	○	○	●
		$\dfrac{1}{4} + \dfrac{2}{3} + \dfrac{7}{8} = \dfrac{1}{4} + \dfrac{2}{3} + \dfrac{7}{8}$	●	○	○
		$\dfrac{6}{7} + \left(-\dfrac{6}{7}\right) = 0$	○	●	○

The associative property of addition means that you can move grouping symbols and it does not change the value of the expression. The inverse property of addition means that you can add a number and its inverse together and you will get 0. The identity property of addition states that if you add 0 to a number it will not change the value.

Lesson 3: Additive Inverse and Distance Between Two Points on a Number Line

Question No.	Answer	Detailed Explanations
1	A	Both terms inside of the parentheses are being subtracted, and subtraction of a number is the same as addition of its opposite.
2	C	The distance between two points can be found by taking the absolute value of the difference of the two points.
3	B	Because Kyle and Mark traveled in opposite directions, let one of their distances be a negative value. The distance between the two of them is the absolute value of the difference of their positions, which is 8 when evaluated.
4	D	The form given is the absolute value of the difference of two numbers. Those two numbers could represent the coordinates of two points if the expression is the distance between them. Since G is located at 5, so H should be at -2.
5	B	Subtracting 3y is the same as adding the opposite of 3y: + (−3y)
6	A	Because they live in opposite directions from the park, let one of their distances from the park be a negative value. The distance between their homes, then, is the absolute value of the difference of the two distances.
7	D	The statement says that subtracting a number is the same as adding the opposite of that number. This is true for all real numbers.
8	C	The x in the expression is being added, not subtracted, so we do not want to replace it with addition of its opposite. The 3 inside parentheses is being subtracted, so we can replace it with addition of its opposite, −3.
9	D	The distance between two points can be represented in the form of the absolute value of the difference of the coordinates of the two points.
10	A	The distance between two points on a number line can always be represented in the form of the absolute value of the difference of the coordinates of the two points.
11	2	If you use the number line to count, the distance between $-1\frac{1}{6}$ and $\frac{5}{6}$ is $\frac{12}{6}$. If you simplify this, you get 2.

Question No.	Answer	Detailed Explanations
12	B and D	B and D are correct options. The opposite of 34 is -34. An elevator descending and removing books represents a negative number. Therefore those are the correct answers.

13		Additive Inverse is added	Additive Inverse is not added
	$3 + (-3)$	●	○
	$-2.2 + 2.2$	●	○
	$1 + 1$	○	●
	$-\frac{4}{7} + \left(-\frac{4}{7}\right)$	○	●

Additive inverse of a number is the opposite of the number. When it is added to the number, the sum becomes zero.

Lesson 4: Converting Between Rational Numbers and Decimals

Question No.	Answer	Detailed Explanations
1	C	Divide 7 by 8 using long division, and the result is 0.875.
2	A	Divide 5 by 6 using long division. The decimal repeats endlessly: 0.83333...
3	A	If the decimal terminates, then the number can always be written as the quotient of two integers and is rational.
4	A	Divide 4 by 11 using long division. The decimal repeats endlessly: 0.363636...
5	B	If the decimal repeats endlessly, the number can be expressed as a quotient of integers and is rational.
6	B	The whole number part is the value of the ones digit. Divide the fraction to learn the decimal part of the number. The result is a repeating decimal: 2.2222...
7	B	Any decimal number that continues endlessly without repeating is not a rational number.
8	D	Divide the numerator by the denominator using long division. The result is 0.44.
9	C	Divide the number of laps by 3 to find how many miles she ran. The result is a repeating decimal: 4.6666...
10	B	Divide the numerator by the denominator using long division. The result is a repeating decimal: 0.343434...
11	0.25	Ratio of number of magazines to number of nonfiction books = $\frac{30}{120}$ = 0.25.
12	A and B	A and B are correct options.

Question No.	Answer	Detailed Explanations
13		

	Repeating Decimal	Terminating Decimal
$\dfrac{3}{4}$		●
$\dfrac{2}{3}$	●	
$\dfrac{1}{9}$	●	
$\dfrac{2}{8}$		●

A decimal is repeating if there is an infinitely repeating sequence of numbers to the right of the decimal point. A decimal is terminating if the digits to the right of the decimal point terminate naturally.

Lesson 5: Solving Real World Problems

Question No.	Answer	Detailed Explanations
1	B	$15.00 - $9.39 = $ 5.61 is the correct answer.
2	B	50.75 x 25 = 1268.75 pounds 1268.75 pounds. Option B is the correct answer.
3	A	1528.80 pounds ÷ 50 bags = 30.576 pounds in each bag. Option A is the correct answer.
4	A	2 bagels @ $0.69 each = 2 x $0.69 = $1.38 3 yogurts @ $1.49 each = 3 x $1.49 = $4.47 1 orange juice @ $1.75 = $1.75 Total bill = $7.60 $7.60 is the correct answer.
5	D	The pizza was divided equally so each person's cost for pizza is $17.49 ÷ 5 = $3.50. Soda was $1.19 each. Add to find the total cost for each person. $3.50 + $1.19 = $4.69. $4.69 is the correct answer.
6	B	1 notebook = $5.69 1 pack paper = $1.39 1 set dividers = $1.45 2 pens = 2 x $1.19 = $2.38 5 pencils = 5 x $0.50 = $2.50 Subtotal = $13.41 Tax = .065 x $13.41 = $0.87 Total = $14.28 The answer is $14.28.
7	B	The lawn is 30 ft x 20 ft = 600 sq ft. He mowed 5 ft x 10 ft = 50 sq ft 600 sq ft - 50 sq ft = 550 sq ft 550 sq ft is the correct answer.
8	C	$\frac{5}{8} \times 5 = \frac{25}{8} = 3\frac{1}{8}$ acres already planted. $5 - 3\frac{1}{8} = 4\frac{8}{8} - 3\frac{1}{8}$ $= 1\frac{7}{8}$ acres remaining to plant. $1\frac{7}{8}$ acres is the correct answer.
9	A	5.25 - 4.98 = 0.27 lb 0.27 lb is the correct answer.
10	B	When multiplying or dividing two integers, like signs will always give a positive, and unlike signs will always give a negative sign. If the signs are both negative, multiply the numbers and give the answer a positive sign, is the correct answer.
11	$1\frac{5}{21}$	$\frac{4}{7} = \frac{(4 \times 3)}{(7 \times 3)} = \frac{12}{21}$ $\frac{2}{3} = \frac{(2 \times 7)}{(3 \times 7)} = \frac{14}{21}$ Adding the two fractions, we get, $\frac{12}{21} + \frac{14}{21} = \frac{26}{21} = 1\frac{5}{21}$

Question No.	Answer	Detailed Explanations
12	A and B	A and B are correct options.

$$\frac{1}{2} = \frac{(1\times5)}{(2\times5)} = \frac{5}{10}$$

$$\frac{3}{5} = \frac{(3\times2)}{(5\times2)} = \frac{6}{10}$$

Adding the two fractions, we get $\frac{5}{10} + \frac{6}{10} = \frac{11}{10} = 1\frac{1}{10}$

13

	Negative	Zero	Positive
$-\frac{6}{7} - \left(-\frac{6}{7}\right)$	○	●	○
$-5 - \frac{3}{5}$	●	○	○
$\frac{3}{4} - \frac{1}{5}$	○	○	●

1. Sum of a number and its additive inverse (i.e. the number with opposite sign) is always zero.

2. Sum of two or more negative numbers will always result in a negative number. In this case we have -5 – 3/5 which can be written as a sum of two negative numbers as (-5) + (-3/5)
= (-5/1 x 5/5) + (-3/5)
=-25/5 + (-3/5)
=-28/5
= -5 3/5

3. ¾ -1/5 = ¾ x 5/5 – 1/5 x 4/4 = 15/20 – 4/20 = 11/20

Lesson 6: Strategies for Adding and Subtracting Rational Numbers

Question No.	Answer	Detailed Explanations
1	A	The commutative property states that parts of an addition statement can be written in any order and the same sum will result.
2	C	The 92 can be written as 82 + 10, so that the 10 and 8 can first be subtracted, and then the result can be added to 82.
3	B	The whole number parts can be added together, and the fraction parts can be added together, resulting in 11 and $\frac{12}{8}$. The $\frac{12}{8}$ part is a full $\frac{8}{8}$, or 1, and an additional $\frac{4}{8}$, so that we have 12 and $\frac{4}{8}$, which simplifies to 12 and $\frac{1}{2}$.
4	C	The distributive property only applies when there is addition or subtraction inside the parentheses and a number outside the parentheses being multiplied to what is inside the parentheses. When you multiply a sum by a number, you can multiply each addend by that number separately and then add the products. It can be represented as follows: a(b + c) = ab + ac.
5	B	The associative property of addition states that different parts of an addition statement can be grouped together in any manner, and the same sum will still result.
6	A	The 7 can be written as 6 + 1, and the 2.45 can be written as (2 + 0.45). Subtract the 2 from the 6 and the 0.45 from the 1.
7	D	$\frac{31}{3}$ is the same as $\frac{30}{3} + \frac{1}{3}$. $\frac{30}{3} = 10$, so we have 10 and $\frac{1}{3}$.
8	B	Let the 16 be 15 + 1, and then subtract the $\frac{7}{8}$ from the 1, leaving only $\frac{1}{8}$. You now have $15 + \frac{1}{8} + \frac{6}{8} - 9$ (Note that $\frac{3}{4}$ is written as $\frac{6}{8}$). Subtract the 9 from the 15, and add the fractions to arrive at 6 and $\frac{7}{8}$. Alternate Method : $16\frac{3}{4} = \frac{67}{4} = \frac{134}{8}$, $9\frac{7}{8} = \frac{79}{8}$ Therefore, $16\frac{3}{4} - 9\frac{7}{8} = \frac{134}{8} - \frac{79}{8} = \frac{134 - 79}{8} = \frac{55}{8} = 6\frac{7}{8}$.
9	A	The identity property of addition states that if you add 0 to a value, the sum is the same as the value itself.

Question No.	Answer	Detailed Explanations
10	C	Let $\frac{20}{7}$ be $\frac{13}{7} + \frac{7}{7}$. Subtract the $\frac{3}{7}$ from the $\frac{13}{7}$, and let the $\frac{7}{7}$ be written as 1 and subtracted with the existing 1.
11	A and D	A and D are correct answers. -26.3-(-12.6) = 13.7 and -3.2 - 10.5 = -13.7

12				

Distance between 4 and -1	$\|-1-4\|$	$\|4-(-1)\|$
Distance between -4 and -1	$\|-1-(-4)\|$	$\|-4-(-1)\|$
Distance between 4 and 1	$\|1-4\|$	$\|4-1\|$
Distance between -4 and 1	$\|1-(-4)\|$	$\|-4-1\|$

The distance between two numbers is the absolute value of the difference between two numbers. It does not matter which number you place first because you are taking the absolute value of the difference.

Lesson 7: Strategies for Multiplying and Dividing Rational Numbers

Question No.	Answer	Detailed Explanations
1	B	When we divide a fraction by another fraction, flip the second fraction (interchange numerator and denominator) and then multiply with the first. $(-\frac{4}{23})$ divided by $(\frac{7}{58}) = (-\frac{4}{23})$ x $(\frac{58}{7})$.
2	C	The commutative property states that multiplication can be done in any order and the same product will result.
3	B	Dividing by a fraction is the same thing as multiplying by the reciprocal of that fraction. Multiply, then, by 2 rather than dividing by $\frac{1}{2}$: 4 x 2 = 8.
4	A	The 3 must be multiplied by all three place values. The same product will result if writing it out in individual terms.
5	A	To find an equivalent expression, change the division symbol to a multiplication symbol and rewrite the second number as a reciprocal. $(\frac{7}{8})$ x $(\frac{15}{1}) = (\frac{7}{8})$ x 15.
6	A	The associative property of multiplication states that multiplication can be grouped in any way and the same product will result.
7	C	Multiply the numerators together and the denominators together. This results in a value of $\frac{21}{12}$, which simplifies to $\frac{7}{4}$.
8	D	Instead of dividing by $\frac{3}{25}$, multiply by the reciprocal, $\frac{25}{3}$. The result is $\frac{150}{15}$, which simplifies to 10.
9	A	Convert both mixed numbers to improper fractions, and multiply them: $(\frac{7}{3})(\frac{7}{2}) = \frac{49}{6}$.
10	B	Convert both fractions to *rational* numbers $(\frac{21}{4}) \div (\frac{3}{2})$ and multiply by the reciprocal $(\frac{21}{4}) \cdot (\frac{2}{3})$ equals $\frac{42}{12}$. $\frac{42}{12}$ reduces to $\frac{7}{2}$.

Question No.	Answer	Detailed Explanations

| 11 | | |

		$\dfrac{6}{4}$	$\dfrac{4}{6}$
		$\dfrac{6}{-19}$	$\dfrac{-19}{6}$
		$\dfrac{5}{-64}$	$\dfrac{-64}{5}$
		$\dfrac{1}{4}$	4

The reciprocal of a fraction or quotient is created by flipping the numerator and denominator.

Lesson 8: Rational Numbers as Quotients of Integers

Question No.	Answer	Detailed Explanations
1	A	0 cannot be the divisor in a division problem. 0 cannot be the divisor in a division problem because division by 0 is undefined.
2	B	Dividing a negative number by a negative number results in a positive number. If only one of the dividend or divisor is negative, then the quotient is negative.
3	C	Divide the length of the trail by the length of one leg, and the number of legs results: 8.
4	A	Dividing the total time Greg ran by the time per mile will result in the total number of miles Greg ran.
5	D	The divisor in a division problem cannot be 0.
6	A	Divide the total amount of water that must be pumped into the pool by the number of gallons per hour, and the result will be the number of hours required to fill the pool.
7	C	In the fraction $\frac{-t}{4}$, the numerator is the opposite of t, and the denominator is the opposite of v. The fact that both numerator and denominator are opposites will result in a cancelling of the two negatives, and the value will be the same as the original. Alternate Explanation : Substitute the value of t = -2 and v = -4 in the options and compare with $\frac{t}{v} = \frac{-2}{-4} = \frac{1}{2}$. (A) $\frac{t}{4} = \frac{-2}{4} = -\frac{1}{2}$ (B) $\frac{4}{2v} = \frac{4}{2 \times -4} = -\frac{1}{2}$ (C) $-\frac{t}{4} = \frac{2}{4} = \frac{1}{2}$ (D) $\frac{-6}{-3v} = \frac{2}{v} = \frac{2}{-4} = -\frac{2}{4} = -\frac{1}{2}$ Therefore, Option (C) is correct.
8	A	We can see that $\frac{a}{b}$ must have a value of 4 in order for $\frac{3a}{b}$ to have a value of 12. The opposite of $\frac{a}{b}$, must be –4.
9	B	The definition of a rational number is that it can be expressed as a ratio (or quotient) of integers.
10	C	Divide the total length in inches (10 feet x 12 inches = 120 inches) by the number of inches per stone, and the quotient will be the number of stones needed. So, 120 inches ÷ 8 inches = 15 inches.
11	-15	-15 divided by 3 = -5

Question No.	Answer	Detailed Explanations
12	C, E are F	C, E are F are correct options. Use the rules for division of integers. In general, division of two numbers with the same sign always equal a positive number. If the signs are different, then the quotient is negative.

13

	Positive	Negative	Zero	Undefined
$-50 \div 5$	○	●	○	○
$\dfrac{0}{-4}$	○	○	●	○
$\dfrac{-35}{-7}$	●	○	○	○
$\dfrac{47}{0}$	○	○	○	●

Chapter 2: Proportional Relationships

Lesson 1: Finding Constant of Proportionality

1. **According to the graph, what is the constant of proportionality?**

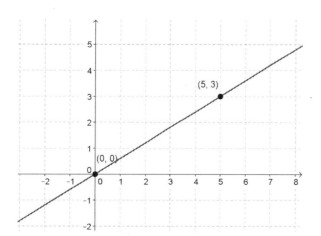

Ⓐ $\dfrac{3}{5}$

Ⓑ 5

Ⓒ 3

Ⓓ $\dfrac{1}{3}$

2. **According to the table, how much does one ticket cost?**

Number of Tickets	Total Cost
3	$ 21.00
4	$ 28.00
5	$ 35.00
6	$ 42.00

Ⓐ $21.00
Ⓑ $4.75
Ⓒ $7.00
Ⓓ $16.50

3. **If y = 3x, what is the constant of proportionality between y and x?**

    Ⓐ 1
    Ⓑ 0.30
    Ⓒ 1.50
    Ⓓ 3

4. **When Frank buys three packs of pens, he knows he has 36 pens. When he buys five packs, he knows he has 60 pens. What is the constant of proportionality between the number of packs and the number of pens?**

    Ⓐ 12
    Ⓑ 10
    Ⓒ 36
    Ⓓ 60

5. **What is the unit rate for the number of hours of study each week per class?**

    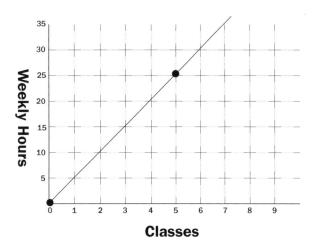

    Classes

    Ⓐ 25 hours
    Ⓑ 5 hours
    Ⓒ 7 hours
    Ⓓ 10 hours

6. **What is the constant of proportionality in the following equation?**

    B = 1.25C

    Ⓐ 1.00
    Ⓑ 0.25
    Ⓒ 1.25
    Ⓓ 2.50

7. **When Georgia buys 3 boxes of peaches, she has 40 pounds more than when she buys 1 box of peaches. How many pounds of peaches are in each box?**

   Ⓐ 40
   Ⓑ 60
   Ⓒ 20
   Ⓓ 3

8. **Look at the table below:**

X	5	6	7	8
Y	90	108	126	144

   What is the constant of proportionality of y to x? Enter your answer in the box given below.

9. **Suppose the relationship between x and y is proportional. If the constant of proportionality of y to x is 16, select possible values for y and x.**

   Ⓐ x = 6 and y = 96
   Ⓑ x = 96 and y = 6
   Ⓒ x = 9 and y = 90
   Ⓓ x = 144 and y = 9

10. In proportional relationships, there are dependent and independent variables. Match each situation to whether it is an independent variable or not.

	Independent Variable	Dependent Variable
The number of concert tickets sold is based on how many hours the ticket booth is open. What kind of variable is the number of hours the ticket booth is open?	☐	☐
The number of cans that is collected for a charity drive is based on how many students bring in cans. What kind of variable is the number of students?	☐	☐
The number of buses needed for a field trip depends on how many students are going on the field trip. What kind of variable is the number of buses?	☐	☐
The number of cell phone towers built is determined by how many customers are in the area. What kind of variable is the number of cell phone towers?	☐	☐

CHAPTER 2 → Lesson 2: Unit Rates

1. **A tennis match was delayed because of rain. The officials were not prepared for the delay. They covered the 25ft by 20ft court with 13ft by 10ft plastic covers. How many plastic covers were needed to cover the court from the rain?**

    Ⓐ 3 plastic covers
    Ⓑ 4 plastic covers
    Ⓒ 5 plastic covers
    Ⓓ 6 plastic covers

2. **John eats a bowl of cereal for 3 of his 4 meals each day. He finishes two gallons of milk in eight days. How much milk does John use for one bowl of cereal? (Assume he only uses the milk for his cereal.)**

    Ⓐ One-twelfth of a gallon of milk
    Ⓑ One cup of milk
    Ⓒ Two cups of milk
    Ⓓ One-sixth of a gallon of milk

3. **A recipe to make a cake calls for three-fourths of a cup of milk. Mary used this cake as the first layer of a wedding cake. The second layer was half the size of the first layer, and the third layer was half the size of the second layer. How much milk would be used for the entire wedding cake?**

    Ⓐ One and two-thirds cups of milk
    Ⓑ One and one-third cups of milk
    Ⓒ One and five-sixteenths cups of milk
    Ⓓ One cup of milk

4. **One third of a quart of paint covers one fourth of a basketball court. How much paint does it take to paint the entire basketball court?**

    Ⓐ one and one-third quarts
    Ⓑ one quart
    Ⓒ one and one-fourth quarts
    Ⓓ one and three-fourths quarts

5. The total cost of 100 pencils purchased at a constant rate is $39.00. What is the unit price?

    (A) $39.00
    (B) $3.90
    (C) $0.39
    (D) $0.039

6. A construction worker was covering the bathroom wall with tiles. He covered three-fifths of the wall with 50 tiles. How many tiles will it take to cover the entire wall?

    (A) 83 tiles
    (B) 83 and one-third tiles
    (C) 85 tiles
    (D) 83 and one-half tiles

7. Jim ran four-fifths of a mile and dropped out of the 1600 meter race. His pace was 12 miles an hour until the point he dropped out of the race. How many minutes did he run?

    (A) 4 minutes and 30 seconds
    (B) 4 minutes
    (C) 4 minutes and 20 seconds
    (D) 4 minutes and 10 seconds

8. Ping played three-fourths of a football game. The game was three and a half hours long. How many hours did Ping play in this game?

    (A) 2 hours 37 minutes
    (B) 2 hours 37 minutes and 30 seconds
    (C) 2 hours 37 minutes and 20 seconds
    (D) 2 hours 37 minutes and 10 seconds

9. Bill is working out by running up and down the steps at the local stadium. He runs a different number of steps in random order.

    Which of the following is his best time of steps per minute?

    (A)   25 steps in 5 minutes
    (B)   30 steps in 5.5 minutes
    (C)   20 steps in 4.5 minutes
    (D)   15 steps in 4 minutes

10. Doogle drove thirty and one-third miles toward his brother's house in one-third of an hour. About how long will the entire hundred mile trip take at this constant speed?

Ⓐ 1 hour
Ⓑ 1 hour and 6 minutes
Ⓒ 1 hour and 1 minutes
Ⓓ 1 hour and 3 minutes

11. A store is selling T-Shirts. Which is the best deal? Select all correct answers that apply.

Ⓐ 8 for $26
Ⓑ 5 for $30
Ⓒ 4 for $15
Ⓓ 12 for $39
    10 for $45

12. Read each sentence and select whether the rate is a rate or unit rate.

	Rate	Unit Rate
The earth rotates 1.25 degrees in 5 minutes.	○	○
Sarah reads 13 pages in 1/3 of an hour.	○	○
A man pays $45.24 for 16 gallons of gasoline.	○	○
The car drives 25 miles per hour.	○	○
The soup costs $1.23 per ounce.	○	○
40 millimeters of rain fell in 1 minute.	○	○

13. What is the unit rate for a pound of seed? Circle the correct answer choice.

Pounds of Seed	Total Cost
10	$17.50
20	$35.00
30	$52.50
40	$70.00

Ⓐ $3.50
Ⓑ $1.75
Ⓒ $17.50
Ⓓ $7.25

CHAPTER 2 → Lesson 3: Applying Ratios and Percents

1. **What value of *x* will make these two expressions equivalent?**

   $$\frac{-3}{7} \text{ and } \frac{x}{21}$$

   Ⓐ x = -3
   Ⓑ x = 7
   Ⓒ x = 9
   Ⓓ x = -9

2. **If *p* varies proportionally to *s*, and *p* = 10 when *s* = 2, which of the following equations correctly models this relationship?**

   Ⓐ p = 5s
   Ⓑ p = 10s
   Ⓒ s = 10p
   Ⓓ 2s = 10p

3. **Solve for x, if** $\dfrac{72}{108}$ **and** $\dfrac{x}{54}$ **are equivalent.**

   Ⓐ x = 18
   Ⓑ x = 36
   Ⓒ x = 54
   Ⓓ x = 24

4. **At one particular store, the sale price, *s*, is always 75% of the displayed price, *d*. Which of the following equations correctly shows how to calculate *s* from *d*?**

   Ⓐ d = 75s
   Ⓑ s = 0.75d
   Ⓒ s = d - 0.75
   Ⓓ s = d + 75

5. **When *x* = 6, *y* = 4. If *y* is proportional to *x*, what is the value for *y* when *x* = 9?**

   Ⓐ   4

   Ⓑ $\dfrac{2}{3}$

   Ⓒ   3
   Ⓓ   6

6. Jim is shopping for a suit to wear to his friend's wedding. He finds the perfect one on sale at 30% off. If the original price was $250.00, what will the selling price be after the discount?

    Ⓐ $75
    Ⓑ $175
    Ⓒ $200
    Ⓓ $220

7. If Julie bought her prom dress on sale at 15% off and paid $110.49 before tax, find the original price of her dress.

    Ⓐ $126.55
    Ⓑ $129.99
    Ⓒ $135.00
    Ⓓ $139.99

8. A plot of land is listed for sale with the following measurements: 1300 ft x 982 ft. When the buyer measured the land, he found that it measured 1285 ft by 982 ft. What was the % of error of the area of the plot?

    Ⓐ 1.47%
    Ⓑ 14.73%
    Ⓒ 1.15%
    Ⓓ 11.7%

9. Pierre received a parking ticket whose cost is $22.00. Each month that he failed to make payment, fees of $7.00 were added. By the time he paid the ticket, his bill was $36.00. What was the ratio of fees to the cost of the ticket?

    Ⓐ $\dfrac{36}{22}$

    Ⓑ $\dfrac{22}{36}$

    Ⓒ $\dfrac{7}{22}$

    Ⓓ $\dfrac{7}{11}$

10. Sara owns a used furniture store. She bought a chest for $42 and sold it for $73.50. What percent did she mark up the chest?

    Ⓐ 100%
    Ⓑ 75%
    Ⓒ 42%
    Ⓓ 31.5%

11. What is 30% of 64? Enter your answer in the box given below.

    [   ]

12. Which of the expressions equals 60%? There can be more than 1 correct answer, select all the correct answers.

    Ⓐ 48 of 100
    Ⓑ 32.4 of 54
    Ⓒ 3 of 5
    Ⓓ 4 of 5
    Ⓔ 32 of 53

13. Read each of the equation and match it to whether it is part, percent or whole.

	Part	Percent	Whole
What is 35% of 15?	☐	☐	☐
4% of what number is 46?	☐	☐	☐
y = .76 X 43	☐	☐	☐
14 = m X 56	☐	☐	☐

CHAPTER 2 → Lesson 4: Understanding and Representing Proportions

1. The following table shows two variables in a proportional relationship:

a	b
2	6
3	9
4	12

   Which of the following is an algebraic statement showing the relationship between a and b?

   Ⓐ a = 3b
   Ⓑ b = 3a
   Ⓒ b = 1/3 (a)
   Ⓓ a = 1/2 (b)

2. If the ratio of the length of a rectangle to its width is 3 to 2, what is the length of a rectangle whose width is 4 inches?

   Ⓐ 4 in.
   Ⓑ 5 in.
   Ⓒ 6 in.
   Ⓓ 7 in.

3. The following table shows two variables in a proportional relationship:

e	f
5	25
6	30
7	35

   Using the relationship between e and f as shown in this table, find the value of f when e = 11.

   Ⓐ 40
   Ⓑ 45
   Ⓒ 50
   Ⓓ 55

4. The following table shows two variables in a proportional relationship:

c	d
4	8
5	10
6	12

If c and d are proportional, then d = kc where k is the constant of proportionality.

Which of the following represents k in this case?

Ⓐ k = 1
Ⓑ k = 2
Ⓒ k = 3
Ⓓ k = 4

5. Ricky's family wants to invite his classroom to a "get acquainted" party. If 20 students attend the party then it will cost $100. Assuming the relationship between cost and guests is proportional, which of the following will be the cost if 29 students attend?

Ⓐ $129
Ⓑ $135
Ⓒ $139
Ⓓ $145

6. Which of the following pairs of ratios form a proportion?

Ⓐ 9 boys to 5 girls and 12 boys to 8 girls
Ⓑ 9 boys to 5 girls and 18 boys to 10 girls
Ⓒ 9 boys to 5 girls and 13 boys to 9 girls
Ⓓ 9 boys to 5 girls and 27 boys to 14 girls

7. If the local supermarket is selling oranges for *p* cents each and Mrs. Jones buys *n* oranges, write an equation for the total *(t)*, that Mrs. Jones pays for oranges.

Ⓐ t = p + n
Ⓑ t = p - n
Ⓒ t = pn
Ⓓ t = p/n

8. **The ratio of Kathy's earnings to her hours worked is constant. This fact implies which of the following?**

Ⓐ Kathy's earnings are proportional to her hours of work.
Ⓑ If Kathy works harder during her 8 hours of work today, her income for today will be greater than if she just takes it easy.
Ⓒ If Kathy only works a half day today, her earnings will be the same as if she worked all day.
Ⓓ If Kathy takes an extra hour for lunch, it will not affect her earnings for the day.

9. **Write a mathematical statement (equation) for the relationship between feet and yards.**

Ⓐ number of feet / number of yards = 3
Ⓑ number of yards / number of feet = 3
Ⓒ number of feet - number of yards = 3
Ⓓ number of feet + number of yards = 3

10. **The ratio of the measurement of a length in yards to the measurement of the same length in feet is a constant. This implies that __?**

Ⓐ the measurement of a length in yards is directly proportional to the measurement of the same length in feet.
Ⓑ the measurement in yards of a length is inversely proportional to the measurement of the same length in feet.
Ⓒ the measurement in yards of a length is equivalent to the measurement of the same length in feet.
Ⓓ There is no relationship between the two measurements.

11. **Look at the table below. Decide whether the table shows a proportional relationship between x and y. Write Yes if the relationship is proportional or write No if the relationship is not proportional in the box given below.**

x	2	4	7	10
y	4	16	49	100

12. **Suppose you are buying pizzas for a party. An equation that represents the cost (y) in dollars for x number of pizzas is y=18x.**

    Does the equation y=18x represent a proportional relationship? Instruction : Check all that are true

    Ⓐ No, the graph of the equation does not pass through the origin.
    Ⓑ Yes, the graph of the equation is a straight line
    Ⓒ Yes, the graph of the equation passes through the origin.
    Ⓓ No, the graph of the equation is not a straight line.

13. **The table shows a proportional relationship between x and y. For each value of x and y, match it to the correct unit rate $\dfrac{y}{x}$ in it's simplest form.**

	$\dfrac{6}{1}$	$\dfrac{3}{1}$	$\dfrac{9}{1}$
x = 3 and y = 27	☐	☐	☐
x = 9 and y = 81	☐	☐	☐
x = 21 and y = 126	☐	☐	☐
x = 14 and y = 84	☐	☐	☐
x = 12 and y = 36	☐	☐	☐
x = 15 and y = 45	☐	☐	☐

CHAPTER 2 → Lesson 5: Represent Proportions by Equations

1. 3 hats cost a total of $18. Which equation describes the total cost, C, in terms of the number of hats, n?

   Ⓐ   C = 3n
   Ⓑ   C = 6n
   Ⓒ   C = 0.5n
   Ⓓ   3C = n

2. Use the data in the table to give an equation to represent the proportional relationship.

x	y
0.5	7
1	14
1.5	21
2	28

   Ⓐ   y = 14x
   Ⓑ   y = 7x
   Ⓒ   7y = x
   Ⓓ   21y = x

3. Kelli has purchased a membership at the gym for the last four months. She has paid the same amount each month, and her total cost so far has been $100. What equation expresses the proportional relationship of the cost and month?

   Ⓐ   C = 100m
   Ⓑ   C = 50m
   Ⓒ   C = 4m
   Ⓓ   C = 25m

4. When buying bananas at the market, Marco pays $4.50 for 5 pounds. What is the relationship between pounds, p, and cost, C?

   Ⓐ   C = 4.5p
   Ⓑ   C = 5p
   Ⓒ   C = 0.9p
   Ⓓ   C = 22.5p

5. The cost to rent an apartment is proportional to the number of square feet in the apartment. An 800 square foot apartment costs $600 per month. What equation represents the relationship between area, a, and cost, C?

   (A) C = 0.75a
   (B) C = 1.33a
   (C) C = 8a
   (D) C = 6a

6. A school has to purchase new desks for their classrooms. They have to purchase 350 new desks, and they pay $7000. What equation demonstrates the relationship between the number of desks, d, and the total cost, C?

   (A) C = 10d
   (B) C = 20d
   (C) C = 70d
   (D) C = 35d

7. A soccer club is hosting a tournament with 12 teams involved. Each team has a set number of players, and there are a total of 180 players involved in the tournament. Which equation represents the proportional relationship between teams, t, and players, p?

   (A) p = 15t
   (B) p = 12t
   (C) p = 18t
   (D) p = 11t

8. A package of three rolls of tape costs $5 per pack. What is the proportional relationship between cost, C, and package of tape, p?

   (A) C = 1.67p
   (B) C = 15p
   (C) C = 3p
   (D) C = 5p

9. Freddy is building a house and has 5 loads of gravel delivered for the work. His total cost for the gravel is $1750. Which equation correctly shows the relationship between cost, C, and loads of gravel, g?

   (A) C = 5g
   (B) C = 350g
   (C) C = 75g
   (D) C = 225g

10. Use the graph to write an equation for the proportional relationship between the number of hours Corrie works, h, and her total pay, P.

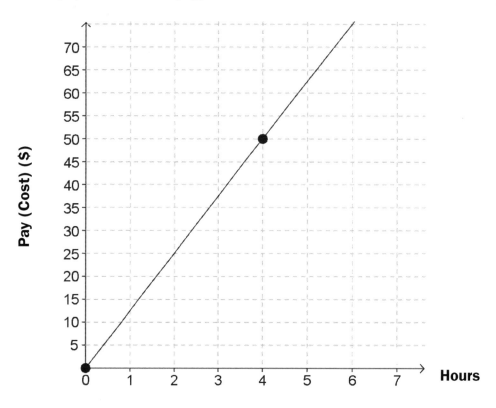

Ⓐ P = 12.50h
Ⓑ P = 50h
Ⓒ P = 4h
Ⓓ P = 22.50h

11. You run 9.1 miles in 1.3 hours at a steady rate. Write an equation that represents the proportional relationship between the x hours you run and the distance y in miles that you travel? Write the equation in the box given below

12. Sally paid $3.50 for 7 apples. Write an equation to represent the total cost y of buying x apples.

Instruction: Select the correct equation to describe the situation. There may be more than one correct answer.

Ⓐ y = 7x

Ⓑ 3.50y = x

Ⓒ y = 0.50x

Ⓓ x = 7

Ⓔ y = 3.50x

Ⓕ $y = \dfrac{3.50}{7}x$

13. Solve each proportion for x and match it with the correct solution for x.

	x = 9	x = 10	x = 7
$\dfrac{x}{15} = \dfrac{3}{5}$	☐	☐	☐
$\dfrac{16}{14} = \dfrac{8}{x}$	☐	☐	☐
$\dfrac{4}{20} = \dfrac{2}{x}$	☐	☐	☐
$\dfrac{75}{x} = \dfrac{50}{6}$	☐	☐	☐

CHAPTER 2 → Lesson 6: Significance of Points on Graphs of Proportions

1. **Which point on the graph of the straight line demonstrates that the line represents a proportion?**

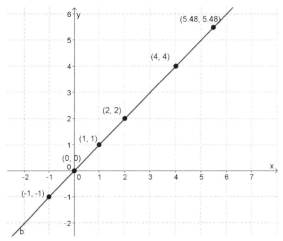

Ⓐ (2, 2)
Ⓑ (0, 0)
Ⓒ (5.48, 5.48)
Ⓓ (-1, -1)

2. **Which point on the graph of the straight line names the unit rate of the proportional relationship?**

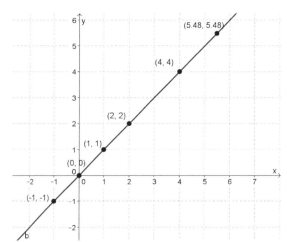

Ⓐ (1, 1)
Ⓑ (0, 0)
Ⓒ (2, 2)
Ⓓ (4, 4)

3. The graph shows the relationship between the number of classes in the school and the total number of students. How many students are there per class?

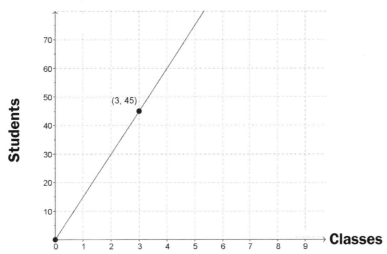

(A) 45 students
(B) 3 students
(C) 135 students
(D) 15 students

4. Use the information given on the graph to find the value of y.

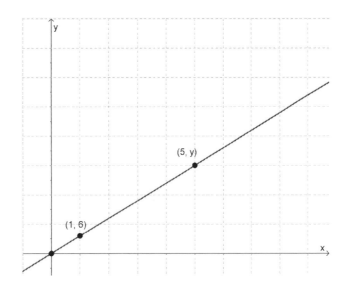

(A) 30
(B) 5
(C) 10
(D) 11

5. In order for the relationship to be proportional, what other point must be a part of the graph?

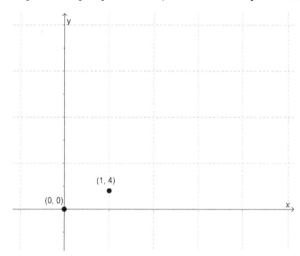

(A) (5, 10)
(B) (5, 25)
(C) (5, 15)
(D) (5, 20)

6. What is the unit rate of Birthday presents per child?

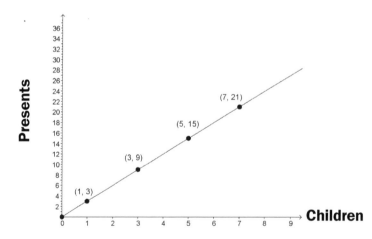

(A) 21 presents
(B) 9 presents
(C) 3 presents
(D) 15 presents

7. **What is the unit rate of bushels of apples per tree?**

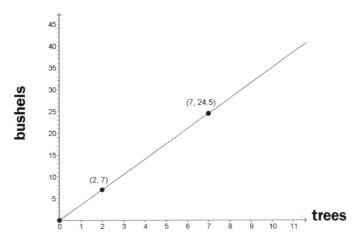

Ⓐ 3.5
Ⓑ 7
Ⓒ 24.5
Ⓓ 14

8. **Which point represents the profit if no boxes of popcorn are sold for the fundraiser?**

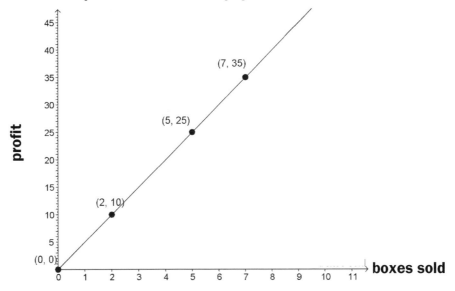

Ⓐ (5, 25)
Ⓑ (7, 35)
Ⓒ (0, 0)
Ⓓ (2, 10)

9. If the relationship between x and y is proportional, what point on the line will indicate the unit rate of that relationship?

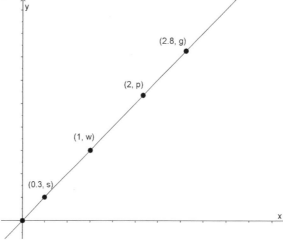

Ⓐ (2, p)
Ⓑ (0.3, s)
Ⓒ (2.8, g)
Ⓓ (1, w)

10. According to the information given on the graph, how much would 20 boards cost?

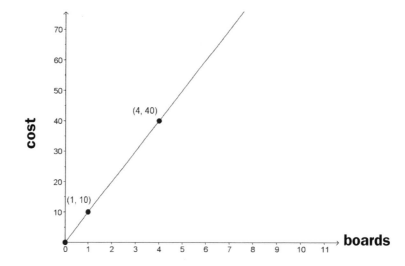

Ⓐ $200
Ⓑ $400
Ⓒ $80
Ⓓ $120

11. Does this graph represent a proportional relationship? Enter your answer as "Yes" or "No" in the box given below.

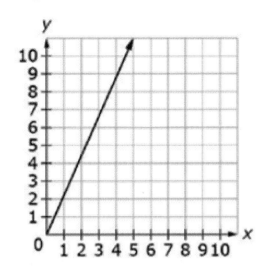

12. What do the points on the graph mean? Select all the correct answers.

Ⓐ After 2 weeks, you have saved $94
Ⓑ After 94 weeks, you have saved $2
Ⓒ After 10 weeks, you have saved $94
Ⓓ After 4 weeks, you have saved $188
Ⓔ After 0 weeks, you have saved $10

13. Match the situations to the corresponding graphs.

Jake ran 3 km in 9 minutes.	○	○
Sara walked 3 feet in 30 seconds.	○	○
After 1 minute, the turtle had traveled 6 feet.	○	○
It took Roger 15 minutes to walk 5 km.	○	○

14. **Plot the points (0,0) and (1,4) on the graph. In order for the relationship to be proportional, what other point must be a part of the graph.**

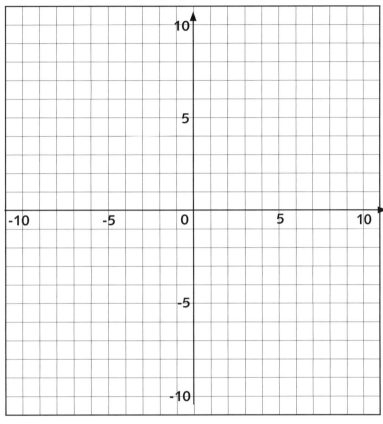

End of Proportional Relationships

Chapter 2: Proportional Relationships

Lesson 1: Finding Constant of Proportionality

Question No.	Answer	Detailed Explanations
1	A	The line passes through the origin, so it is a proportional relationship. The second point gives the constant of proportionality, which is the y value divided by the x value: $\frac{3}{5}$.
2	C	For any of the pairs of data, divide the total cost by the number of tickets to find how much one ticket costs. For example, $\frac{\$21}{3} = \$7$ per ticket.
3	D	Because the equation is solved for y and is equal to a multiple of x, the coefficient of the x term gives the constant of proportionality: 3.
4	A	Divide the number of pens by the number of packs for either pair of data. For example, $\frac{36}{3} = 12$ pens per pack.
5	B	The line passes through the origin, so it is a proportional relationship. The second point gives the constant of proportionality, which is the y value divided by the x value: $\frac{25}{5}$, which simplifies to 5.
6	C	Because the equation is solved for y and is equal to a multiple of x, the coefficient of the x term gives the constant of proportionality: 1.25.
7	C	The difference between our two pairs of data is 2 boxes and 40 pounds of peaches. Divide the difference in pounds by the difference in boxes: $\frac{40}{2} = 20$ pounds per box.

Question No.	Answer	Detailed Explanations
8	18	The constant of proportionality is 18 because the relationship between x and y is to multiply by 18. For example, 5x18=90, 6x18=108, 7x18=126, 8x18=144. You can find this by dividing y by x.
9	A	The constant of proportionality of y to x is determined by dividing y by x. Because $\frac{96}{6} = 16$.
10		The number of hours the ticket booth is open is an independent variable because the number of tickets sold is dependent on the number of hours the ticket booth is open. The number of cell phone towers built is the dependent variable because the number of cell phone towers built is determined by how many customers are in the area. The number of buses is the dependent Variable because the number of buses depends on how many students are going on the field trip. The number of students who bring in cans is the independent variable because the number of cans collected is based on how many students bring in cans.

Lesson 2: Unit Rates

Question No.	Answer	Detailed Explanations
1	B	First, find out the area of the tennis court. (1) 25 × 20 = 500 square feet Next, find out the area of one plastic cover (2) 13 × 10 = 130 square feet Divide 500 by 130 (3) 500 ÷ 130 = 3.84 Therefore, 4 plastic covers are needed for the tennis court.
2	A	First, find out how much cereal John eats in 8 days: (1) 3 bowls per day x 8 days = 24 bowls. Since it takes 2 gallons of milk to eat 24 bowls of cereal, set up the ratio and simplify: (2) $\frac{2}{24}$(GCF is 2, so divide numerator and denominator to find simplest form) (3) $\frac{1}{12}$ Therefore, John uses $\frac{1}{12}$ of a gallon of milk in each bowl of cereal.
3	C	To solve this problem, first, find the amount of milk required for each layer. For the first layer, $\frac{3}{4}$ cup of milk is required. Since the second layer is half the size, the amount of milk required will be $\frac{3}{4} \times \frac{1}{2} = \frac{3}{8}$ cups are required. For the third layer, which is half of layer two, the milk required will be $\frac{3}{8} \times \frac{1}{2} = \frac{3}{16}$ of a cup. The total milk required will be $\frac{3}{4} + \frac{3}{8} + \frac{3}{16}$ To add, find the LCD, which is 16 in this case. Hence, $\frac{3}{4} + \frac{3}{8} + \frac{3}{16}$ represented by LCD will become $(\frac{3}{4} \times \frac{4}{4}) + (\frac{3}{8} \times \frac{2}{2}) + \frac{3}{16}$ $= \frac{12}{16} + \frac{6}{16} + \frac{3}{16} = \frac{21}{16}$ Converting this to mixed fraction we get, $1\frac{5}{16}$. Hence, the correct answer choice is Option C.

Question No.	Answer	Detailed Explanations
4	A	If $\frac{1}{4}$ th of a basketball court can be covered by $\frac{1}{3}$ rd of a quart of paint then to cover the entire court 4 times the amount required for covering $\frac{1}{4}$ th of a court would be required. Therefore, Paint required to cover the entire court = $\frac{1}{3} \times 4 = \frac{4}{3} = 1\frac{1}{3}$
5	C	Unit price means price per one unit. Therefore, we need to know the price per pencil. Since the price of the pencils is defined to be a constant rate, then the total cost ($39.00) divided by the total number of pencils (100) will give us the cost per pencil (or per unit). $39.00 / 100 = $0.39 per pencil The unit price is $0.39. The correct answer is $0.39.
6	B	To solve this problem, find out how much it takes to cover 1/5 of the wall, and then multiply by 5. (1) 50 ÷ 3 = (how much it takes to cover 1/5 of the bathroom wall) (2) multiply 16.66 × 5 = 83.33 (3) 83.33 = $83\frac{1}{3}$ Therefore, it will take 83 and one-third tiles to cover the entire wall.
7	B	To solve this problem, set up a proportion: (distance)/(time) = (distance)/(time) Plug in the numbers (you can convert to decimals to simplify the process): $\frac{0.8 \text{ mile}}{x \text{ min}} = \frac{12 \text{ miles}}{60 \text{ min}}$. Using cross products a/b = c/d is ad = bc (0.8)(60) = (x)(12) 48 = 12x (solve for x by dividing each side by 12)4 = x. Therefore, Jim ran 4 minutes.
8	B	To solve this problem, multiply $\frac{3}{4} \times 3\frac{1}{2}$ (1). Convert $3\frac{1}{2}$ to an improper fraction $= \frac{7}{2}$ (2) $\frac{3}{4} \times \frac{7}{2}$ = (3) $\frac{3x7}{4x2} = $ (4) $\frac{21}{8}$ (convert back to a mixed number)(5) $2\frac{5}{8}$ (6) $\frac{5}{8}$ = 37.5 minutes. Therefore, Ping played 2 hours 37 minutes and 30 seconds.

Question No.	Answer	Detailed Explanations
9	B	To answer the question we must find in which case he ran the most steps per minute. Since he is running up and down, we double the number of steps; so $\dfrac{50}{5} = 10$ steps per minute $\dfrac{60}{5.5} = 10.9$ steps per minute * (when rounded to the nearest tenth) $\dfrac{40}{4.5} = 8.9$ steps per minute (when rounded to the nearest tenth) $\dfrac{30}{4} = 7.5$ steps per minute 30 steps in 5.5 minutes gives an average of 10.9 steps per minute which is his best time. 30 steps in 5.5 minutes is the correct answer. (Note : If we don't double the steps, then also answer does not change. Only the number of steps per minute will differ)
10	B	To solve this problem, set up a proportion: (distance)/(time) = (distance)/(time) $\dfrac{1}{3}$ of an hour is equivalent to 20 minutes. Plug in the numbers (you can convert to decimals to simplify the process): $\dfrac{30.33 \text{ miles}}{20 \text{ minutes}} = \dfrac{100 \text{ miles}}{x \text{ minutes}}$. Using cross products a/b = c/d is ad = bc $(30.33)(x) = (20)(100)$ $30.33x = 2000$ (solve for x by dividing each side by 30.33) x = 65.94 (round up to the nearest whole number). Since 65.94 is almost 66 minutes, that is the same as one hour and 6 minutes.
11	A and D	A and D are correct options. $26 divided by 8 = $3.25 $39 divided by 12 = $3.25 $30 divided by 5 = $6 $15 divided by 4 = $3.75 $45 divided by 10 = $4.50 So the best deal is the deal where the unit cost is the lowest. Therefore the two deals where the unit rates are $3.25 is the correct answer.

Question No.	Answer	Detailed Explanations

12

	Rate	Unit Rate
The earth rotates 1.25 degrees in 5 minutes.	●	○
Sarah reads 13 pages in 1/3 of an hour.	●	○
A man pays $45.24 for 16 gallons of gasoline.	●	○
The car drives 25 miles per hour.	○	●
The soup costs $1.23 per ounce.	○	●
40 millimeters of rain fell in 1 minute.	○	●

The unit rates are rates that reflect amounts per 1 unit. Rates are comparisons of two amounts that are not zero.

13 — **B** — For any of the pairs of data, divide the total cost by the number of pounds of seed to find how much one pound costs. For example, $17.50/10 = $1.75 per pound.

Lesson 3: Applying Ratios and Percents

Question No.	Answer	Detailed Explanations
1	D	In a set of equivalent ratios, or a proportion, the numerator and denominator of one ratio will be multiplied by the same number to get the values of the other ratio. In this case, the denominator of the first ratio, 7, is multiplied by 3 to get to 21. This means (-3) must also be multiplied by 3 to get to (-9).
2	A	To find the constant of proportionality, find the relationship between p and s. When p = 10 and s = 2, dividing p by s shows that p is 5 times s. Therefore, the equation that shows the constant of proportionality is p = 5s.
3	B	To solve this proportion for x, multiply both sides of the equation by 54, and simplify the result, We get, x = 36.
4	B	To find the constant of proportionality, find the relationship between s and d. s is 75% of d, which is the same as 0.75 times d. Therefore, the equation that shows the constant of proportionality is s = 0.75d.
5	D	Set up a proportion between the known ratio and the unknown ratio, and solve for y. $\dfrac{6}{4} = \dfrac{9}{y}$ $6y = 9\,(4)$ — Cross Multiply $6y = 36$ — Simplify $\dfrac{6y}{6} = \dfrac{36}{6}$ — Divide each side by 6 $y = 6$ — Simplify
6	B	Selling price (sp) = Original price (op) - 0.30(op) sp = 250 - 0.30(250) sp = 250 - 75 sp = 175. $175 is the correct answer.
7	B	Original price (op) - 0.15(op) = Selling price (sp) op - 0.15(op) = 110.49 0.85(op) = 110.49 $op = \dfrac{110.49}{0.85}$ op = $129.99. $129.99 is the correct answer.

Question No.	Answer	Detailed Explanations

8 — **C**

Using original measurements, Area=1300 x 982=1,276,600 sq ft.
Using actual measurements, Area=1285 x 982=1,261,870 sq ft.
Error = 14,730 sq ft.

$$\frac{14,730}{1,276,600} \times 100 = 1.15\% \text{ error.}$$

The correct answer is 1.15%.

Alternate Explanation : Since width is same in both the cases, we can

write error $=\dfrac{\text{(original length x width) - (measured length x width)}}{\text{(original length x width)}}$

$=\dfrac{(1300 \times 982) - (1285 \times 982)}{(1300 \times 982)} = 982$

$\dfrac{1300 - 1285}{1300 \times 982}$. Since 982 is common factor in both numerator and

denominator, error $=\dfrac{1300 - 1285}{1300} = \dfrac{15}{1300} = 0.0115 = 1.15\%$

9 — **D**

$36.00 - $22.00 = $14.00 in fees. $\dfrac{\$14.00}{\$22.00} = \dfrac{14}{22} = \dfrac{7}{11}$ is the

ratio of fees to the cost of the ticket. $\dfrac{7}{11}$ is the correct answer.

10 — **B**

Selling price - cost = markup
$73.50 - $42.00 = $31.50 markup
markup / (cost - percent of markup)

$\dfrac{\$31.50}{\$42} = 0.75 = 75\%$ markup. The correct answer is 75%.

11 — **19.2**

64 times 0.30=19.2. Use the percent equation: part=percent times whole.

12 — **B and C**

B and C are correct options. 60% of 54 is 32.4 (0.60*54=32.4)
60% of 5 is 3 (0.60*5=3)

13

	Part	Percent	Whole
What is 35% of 15?	☐		
4% of what number is 46?			☐
y = .76 × 43	☐		
14 = m × 56	☐		

Use the percent equation to solve. Part = percent times whole. The word "what" tells you what the question is looking for.

Lesson 4: Understanding and Representing Proportions

Question No.	Answer	Detailed Explanations
1	B	In each case, if we multiply a by 3, we get b. b = 3a is the correct answer.
2	C	We know that, length/width = $\frac{3}{2}$ The width is given as 4 inches, Hence, we get, $\frac{3}{2} = \frac{x}{4}$ where x is the length. Cross multiplying we get, $x = \frac{3 \times 4}{2}$ $x = \frac{12}{2} = 6$ in Hence, Option C is the correct answer choice.
3	D	The table shows that f = 5e, then 5 is the constant of proportionality. Therefore, f = 5e; so if e = 11, f = 5 x 11 = 55. Hence, 55 is the correct answer.
4	B	In each case, d = 2c. Therefore, k = 2. 2 is the constant of proportionality. k = 2 is the correct answer.
5	D	Since the relationship is proportional we can write the mathematical statement, $\frac{20}{100}$ = 29/c. So 20c = 2900. c = \$145. \$145 is the correct answer.
6	B	A proportion consists of two equivalent ratios. $\frac{9}{5} = \frac{18}{10}$ because 9 x 10 = 5 x 18. 9 boys to 5 girls and 18 boys to 10 girls is the correct answer.
7	C	In this case, p is constant regardless of the number, n, of oranges that she buys. To find the total cost, t, we multiply the number of oranges purchased, n, by the cost per orange, p. t = pn is the correct answer.
8	A	If Kathy's earnings are proportional to her hours of work, then the ratio of her earnings to hours worked is a constant. "Kathy's earnings are proportional to her hours of work" is the correct answer.
9	A	Since 3 ft = 1 yd, if we divide feet by yards, the quotient will always be 3. "number of feet / number of yards = 3" is the correct answer.
10	A	Since these ratios are always equivalent to a constant, the variables are directly proportional by definition. "The measurement in yards of a length is directly proportional to the measurement of the same length in feet" is the correct answer.

Question No.	Answer	Detailed Explanations
11	No	The table shows a relationship that is not proportional because the relationship between x and y is not constant. For example, when x=2 and y=4, the constant of proportionality is 2. However, when x=4 and y=16, the constant of proportionality is 4. There is a relationship between x and y but it is not proportional.
12	B and C	B and C are correct options. One can check whether or not an equation of a line represents a proportional relationship by observing whether the equation represents a straight line and whether the line passes through the origin. The correct answer is that yes, the equation represents a proportional relationship because the graph is a straight line and also because it passes through the origin.
13		

	$\dfrac{6}{1}$	$\dfrac{3}{1}$	$\dfrac{9}{1}$
x = 3 and y = 27	☐	☐	■
x = 9 and y = 81	☐	☐	■
x = 21 and y = 126	■	☐	☐
x = 14 and y = 84	■	☐	☐
x = 12 and y = 36	☐	■	☐
x = 15 and y = 45	☐	■	☐

To determine the ratio $\dfrac{Y}{X}$ you should divide y by x or simplify the fraction to simplest form. Each of the items match only one of the three ratios at the top of the table.

Lesson 5: Represent Proportions by Equations

Question No.	Answer	Detailed Explanations
1	B	Divide the total cost by the number of hats: $\frac{18}{3} = 6$. This gives the cost per hat, which is the constant of proportionality in the equation: C = 6n.
2	A	Choose a pair of data. Divide the value of y by the value of x: $\frac{14}{1} = 14$. This gives the constant of proportionality in the equation: y = 14x.
3	D	Divide the total cost by the number of months: $\frac{100}{4} = 25$. This gives the cost per month, which is the constant of proportionality in the equation: C = 25m.
4	C	Divide the total cost by the number of pounds: $\frac{4.50}{5} = 0.9$. This gives the cost per pound, which is the constant of proportionality in the equation: C = 0.9p.
5	A	Divide the rental cost by the area of the apartment: $\frac{600}{800} = 0.75$. This gives the cost per square foot of area, which is the constant of proportionality in the equation: C = 0.75a.
6	B	Divide the total cost by the number of desks: $\frac{7000}{350} = 20$. This gives the cost per desk, which is the constant of proportionality in the equation: C = 20d.
7	A	Divide the total number of players by the number of teams: $\frac{180}{12} = 15$. This gives the number of players per team, which is the constant of proportionality in the equation: p = 15t.
8	D	The relationship named is that between cost and the number of packs. The number of rolls per pack is irrelevant. The cost per pack is already given as $5 per pack, so this is the constant of proportionality in the equation: C = 5p
9	B	Divide the total cost by the number of loads: $\frac{1750}{5} = 350$. This gives the cost per load, which is the constant of proportionality in the equation: C = 350g.

Question No.	Answer	Detailed Explanations
10	A	The straight line passes through the origin, so the relationship is proportional. Use the values of a point on the line. Divide the pay by the number of hours: $\frac{50}{4}$ = 12.50. This gives the pay per hour, which is the constant of proportionality in the equation: P = 12.50h.
11	y=7x	To solve this problem you need to find the constant of proportionality between x and y. You do this by dividing 9.1 by 1.3 and that equals 7. Therefore, y = 7x. Another way of looking at this is finding the unit rate by dividing 9.1 by 1.3. The unit rate is 7 miles per one hour. That translates into the equation y=7x.
12	C and F	C and F are the correct options. First you find the unit rate by dividing 3.50 by 7. The unit rate is 0.50. To find the cost of buying x number of apples, you would multiply 0.50 by x.

Question No. 13

	x = 9	x = 10	x = 7
$\frac{x}{15} = \frac{3}{5}$	■	☐	☐
$\frac{16}{14} = \frac{8}{x}$	☐	☐	■
$\frac{4}{20} = \frac{2}{x}$	☐	■	☐
$\frac{75}{x} = \frac{50}{6}$	■	☐	☐

To solve for x, you calculate the cross products and solve for x. Therefore, 15 times 3=5x, 75 times 6=50x, 14 times 8 = 16x, and 20 times 2 = 4x. Solve for x by dividing by the coefficient next to x.

Lesson 6: Significance of Points on Graphs of Proportions

Question No.	Answer	Detailed Explanations
1	B	A straight line graph is a proportion if and only if it passes through the origin, (0, 0).
2	A	The unit rate is the amount per single unit of something. This means that the point located at 1 on the x-axis will indicate the unit rate: (1, 1).
3	D	Find the unit rate, which is the number of students per one class, by dividing the y value by the x value for the given point: $\frac{45}{3} = 15$ students per class.
4	A	The unit rate shows that y is 6 times the value of x. Multiply the x value of 5 by 6, and 30 is the value of y.
5	D	The unit rate is given by the point (1, 4). y must be 4 times x. That relationship is found only in the point (5, 20).
6	C	The unit rate is found at the point where x has a value of 1. That point is (1, 3), so the unit rate is 3 presents per child.
7	A	The unit rate can be found by dividing the y value by the x value for any given point: $\frac{7}{2} = 3.5$ bushels per tree.
8	C	The point (0, 0) always represents the origin of the proportion. In this case, that means that no boxes are sold, and the profit will be $0.
9	D	The unit rate is always located at the point where the x value is 1. In this case, that is (1, w).
10	A	The unit rate is found at (1, 10), which indicates that 1 board has a cost of $10. Multiply by 20 boards to find the total cost: (10)(20) = $200.
11	Yes	Yes it does because it is a straight line that passes through the origin.
12	A and D	A and D are the correct options. You use the x and y coordinates to determine the meaning of the points. The point (2,94) means that after two weeks you have saved $94. The point (4,188) means that after 4 weeks you have saved $188.

Question No.	Answer	Detailed Explanations

13

Jake ran 3 km in 9 minutes.	●	○
Sara walked 3 feet in 30 seconds.	○	●
After 1 minute, the turtle had traveled 6 feet.	○	●
It took Roger 15 minutes to walk 5 km.	●	○

Jake ran 3 km in 9 minutes describes the point (9,3). Sara walked 3 feet in 30 seconds describes the point (0.5,3). After 1 minute, the turtle had traveled 6 feet describes the point (1,6). It took Roger 15 minutes to walk 5 km describes (15,5).

14

Coordinates are represented as (x,y)
(0,0) means 0 on x axis and 0 on y axis
(1,4) means 1 on x axis and 4 on y axis.

Plot the points.
The unit rate is given by the point (1, 4). y must be 4 times x, y = 4x. That relationship is found in points (2,8) , (3,12), (4,16), (5,20) etc.,

Chapter 3: Algebra

Lesson 1: Applying Properties to Rational Expressions

1. Ruby is two years younger than her brother. If Ruby's brother's age is A, which of the following expressions correctly represents Ruby's age?

   Ⓐ A - 2
   Ⓑ A + 2
   Ⓒ 2A
   Ⓓ 2 - A

2. Find the difference: 8n - (3n - 6) =

   Ⓐ -n
   Ⓑ 5n - 6
   Ⓒ 5n + 6
   Ⓓ 8n - 6

3. Find the sum:

   6t + (3t - 5) =

   Ⓐ 9t - 5
   Ⓑ 9t + 5
   Ⓒ 3t - 5
   Ⓓ 6t - 5

4. Combine like terms and factor the following expression.

   7x - 14x + 21x - 2

   Ⓐ 15x - 2
   Ⓑ 2(7x - 1)
   Ⓒ 42x - 2
   Ⓓ 21(x - 1)

5. Which of the following expressions is equivalent to:

   3(x + 4) - 2

   Ⓐ 3x + 10
   Ⓑ 3x + 14
   Ⓒ 3x + 4
   Ⓓ 3x + 5

6. **Simplify the following expression:**

$$\left(\frac{1}{2}\right)x + \left(\frac{3}{2}\right)x$$

Ⓐ 2x

Ⓑ $\left(\dfrac{5}{2}\right)$ x

Ⓒ - x

Ⓓ $\dfrac{x}{2}$

7. **Simplify the following expression:**

0.25x + 3 - 0.5x + 2

Ⓐ -0.25x + 5
Ⓑ 0.75x + 5
Ⓒ -0.25x + 1
Ⓓ 5.75x

8. **Which of the following statements correctly describes this expression?**

2x + 4

Ⓐ Four times a number plus two
Ⓑ Two more than four times a number
Ⓒ Four more than twice a number
Ⓓ Twice a number less four

9. **Which of the following statements correctly describes the following expression?**

$$\frac{2x - 3}{2}$$

Ⓐ Half of three less than twice a number
Ⓑ Half of twice a number
Ⓒ Half of three less than a number
Ⓓ Three less than twice a number

10. **Which of the following expressions is not equivalent to:**

$$(\frac{1}{2})(2x + 4) - 3$$

Ⓐ (x + 2) - 3

Ⓑ $(\frac{1}{2})(2x + 4) - 3$

Ⓒ x - 1

Ⓓ x + 1

11. **Which property is demonstrated in the following expression?**

12(3x - 9) = 36x - 108

Ⓐ Associative property
Ⓑ Distributive property
Ⓒ Identity property of addition
Ⓓ Zero property of multiplication

12. **Rhonda is purchasing fencing to go around a rectangular lot which is 4x + 9 ft long and 3x - 5 ft wide. Which expression represents the amount of fencing she must buy?**

Ⓐ 7x + 4
Ⓑ 7x - 4
Ⓒ 14x + 28
Ⓓ 14x + 8

13. **Rebekah is preparing for a swim meet. She is trying to swim 1 mile in 7 minutes. If the pool is 5x + 3 ft long, which expression represents how many laps she needs to swim in 7 minutes?**

Assume 1 length of the pool is 1 lap. (1mile = 5280 feet.)

Ⓐ 7(5x + 3)
Ⓑ 5x + 3
Ⓒ $\frac{5280}{5x + 3}$
Ⓓ 5280(5x + 3)

14. Simplify.

5x + 10y + 0(z) =

Ⓐ 0
Ⓑ 5x + 10y
Ⓒ 15xy
Ⓓ 5x

15. Which property is demonstrated below?

2 + (8 + 3) = (2 + 8) + 3

Ⓐ Additive Identity Property
Ⓑ Multiplicative Identity Property
Ⓒ Distributive Property
Ⓓ Associative Property of Addition

16. Use the Distributive Property to expand the expression 6(5x – 3). Write your answer in the box given below.

17. Which expressions are equal to 60x - 24? There is more than one correct answer. Select all the correct expressions.

Ⓐ 10(6x - 2)
Ⓑ 4(10x - 6)
Ⓒ 5(12x - 5)
Ⓓ 6(10x - 4)
Ⓔ -6(-10x + 4)
Ⓕ 4(15x - 6)

18. A bookstore is advertising $2 off the price of each book. You decide to buy 8 books. Let p represent the price of each book. Use the expression 8(p – 2) to find out how much you would spend if the regular price of each book is $13.

Write your answer in the box given below.

CHAPTER 3 → Lesson 2: Modeling Using Equations or Inequalities

1. A 30 gallon overhead tank was slowly filled with water through a tap. The amount of water (W, in gallons) that is filled over a period of t hours can be found using W = 3.75(t). If the tap is opened at 7 AM and closed at 3 PM, how much water would be in the tank? Assume that the tank is empty before opening the tap.

   Ⓐ 18 gallons
   Ⓑ 20 gallons
   Ⓒ 24 gallons
   Ⓓ The tank is full

2. The ratio (by volume) of milk to water in a certain solution is 3 to 8. If the total volume of the solution is 187 cubic feet, what is the volume of water in the solution?

   Ⓐ 130 cubic feet
   Ⓑ 132 cubic feet
   Ⓒ 134 cubic feet
   Ⓓ 136 cubic feet

3. A box has a length of 12 inches and width of 10 inches. If the volume of the box is 960 cubic inches, what is its height?

   Ⓐ 6 inches
   Ⓑ 10 inches
   Ⓒ 12 inches
   Ⓓ 8 inches

4. Jan is planting tomato plants in her garden. Last year she planted 24 plants and harvested 12 bushels of tomatoes during the season. This year she has decided to only plant 18 plants. If the number of plants is directly proportional to the number of bushels of tomatoes harvested, how many bushels of tomatoes should she expect this year?

   Ⓐ 6
   Ⓑ 18
   Ⓒ 12
   Ⓓ 9

5. Melanie's age added to Roni's age is 27. Roni's age subtracted from Melanie's age is 3. Find their ages.

   Ⓐ 17, 10
   Ⓑ 16, 11
   Ⓒ 15, 12
   Ⓓ 14, 13

6. Tim wraps presents at a local gift shop. If it takes 2.5 meters of wrapping paper per present, how many can Tim wrap if he has 50 meters of wrapping paper?

 Ⓐ 18
 Ⓑ 20
 Ⓒ 15
 Ⓓ 17

7. Name the property demonstrated by the equation.

 $11 + (8 + 6) = (y + 8) + 6$. Also, find the value of y?

 Ⓐ 11, Commutative Property of Addition
 Ⓑ 11, Associative Property of Addition
 Ⓒ 11, Distributive Property
 Ⓓ 11, Associative Property of Multiplication

8. In the linear equation $p = 2c + 1$, c represents the number of couples attending a certain event, and p represents the number of people at that event.

 If there are 7 couples attending, how many people will be present?

 Ⓐ 7 people
 Ⓑ 14 people
 Ⓒ 15 people
 Ⓓ 16 people

9. The ratio (by volume) of salt to sugar in a certain mixture is 4 to 8. If the total volume of the mixture is 300 cubic feet, what is the volume of sugar in the mixture?

 Ⓐ 200 cubic feet
 Ⓑ 199 cubic feet
 Ⓒ 201 cubic feet
 Ⓓ 136 cubic feet

10. Which of the following sequences follows the rule $(8 + t^2) - 2t$ where t is equal to the number's position in the sequence?

 Ⓐ 7, 8, 11, 16, 23, ...
 Ⓑ 9, 12, 17, 24, 33 ...
 Ⓒ 3, 4, 5, 6, 7, ...
 Ⓓ 7, 10, 15, 22, 31, ...

11. Use the model to solve the equation. Circle the correct answer choice.

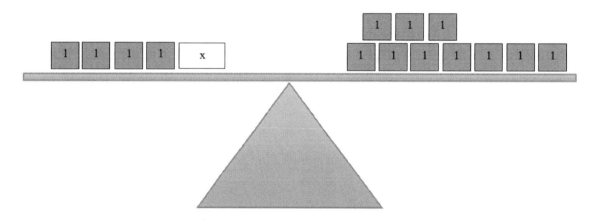

    Ⓐ 6
    Ⓑ 4
    Ⓒ 1
    Ⓓ 10

12. Which equation correctly represents the model and what is the solution?

Note: Select the answers that have the correct equation and solution. There is more than one correct answer.

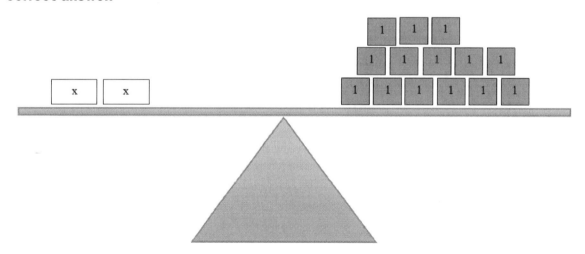

    Ⓐ 2x = 14 and x = 7
    Ⓑ x + 2 = 14 and x = 12
    Ⓒ 2 + x = 14 and x = 12
    Ⓓ x + x = 14 and x = 7
    Ⓔ 2 = 14x and x = 0.5

CHAPTER 3 → Lesson 3: Linear Inequality Word Problems

1. The annual salary for a certain position depends upon the years of experience of the applicant. The base salary is $50,000, and an additional $3,000 is added to that per year of experience, y, in the field. The company does not want to pay more than $70,000 for this position, though. Which of the following inequalities correctly expresses this scenario?

    Ⓐ 53,000y ≤ 70,000
    Ⓑ 3,000y ≤ 50,000
    Ⓒ 50,000 + 3,000y ≤ 70,000
    Ⓓ 3,000 + 50,000y ≤ 70,000

2. Huck has $225 in savings, and he is able to save an additional $45 per week from his work income. He wants to save enough money to have at least $500 in his savings. If w is the number of weeks, express this situation as an inequality.

    Ⓐ 265w ≥ 500
    Ⓑ 225 + 45w ≥ 500
    Ⓒ 225 ≤ 45w
    Ⓓ 225 + 45w ≤ 500

3. Lucy is charging her phone. It has a 20% charge right now and increases by an additional 2% charge every 3 minutes. She doesn't want to take it off the charger until it is at least 75% charged. If m is the number of minutes Lucy keeps her phone for charging, express this situation in an inequality.

    Ⓐ $20 + \dfrac{2}{3} m \leq 75$

    Ⓑ $20m + \dfrac{2}{3} m \leq 75$

    Ⓒ $75 + \dfrac{2}{3} m \geq 20$

    Ⓓ $20 + \dfrac{2}{3} m \geq 75$

4. Matt's final grade, G, in class depends on his last test score, t, according to the expression G = 74 + 0.20t. If he wants to have a final grade of at least 90.0, what is the minimum score he can make on the test?

    Ⓐ 80
    Ⓑ 74
    Ⓒ 92
    Ⓓ 86

5. Students can figure out their grade, G, on the test based upon how many questions they miss, n. The formula for their grade is G = 100 − 4n. If Tim wants to make at least an 83, what is the maximum number of questions he can miss?

    Ⓐ 5 questions
    Ⓑ 3 questions
    Ⓒ 4 questions
    Ⓓ 6 questions

6. The necessary thickness, T, of a steel panel in mm depends on the unsupported length, L, of the panel in feet according to the inequality T ≥ 3 + 0.2L. If the thickest panel available has a thickness of 12 mm, what is the maximum length it can span?

    Ⓐ 75 ft
    Ⓑ 45 ft
    Ⓒ 60 ft
    Ⓓ 54 ft

7. The pay that a salesperson receives each week is represented by the inequality P ≥ 300 + 25s, where s represents the number of units the salesperson sells. What is the significance of the number 300 in this inequality in this context?

    Ⓐ The salesperson will never make more than $300 weekly.
    Ⓑ The salesperson has never sold more than 300 units in a week.
    Ⓒ The salesperson is guaranteed at least $300 even if he doesn't sell any units.
    Ⓓ $300 is the price of a single unit.

8. The net calories gained or lost by a dieter depends on the hours exercised in a week according to the equation C = 750 − 200h. Based on his goals, Neil determines that the following inequality is necessary for him: h > 3.75. What is the significance of the value of this inequality?

    Ⓐ Neil will lose at least 3.75 pounds per week if he meets this inequality.
    Ⓑ Neil will lose weight each week if he meets this inequality.
    Ⓒ Neil can eat only 375 calories on any given day.
    Ⓓ If Neil meets this inequality, he will be able to eat as many calories as he likes.

9. A school always allows twice as many girls as boys to enroll in swimming classes each year. They also follow a certain inequality for the number of boys, B, and girls, G, in their admissions: B + G ≤ 300. What significance does this inequality have for the number of boys that can be admitted each year?

    Ⓐ At least 200 boys must be admitted.
    Ⓑ No more than 100 boys can be admitted.
    Ⓒ The number of boys admitted must equal the number of girls.
    Ⓓ No more than 150 boys can be admitted.

10. Isaac has to do at least 4 hours of chores each week. For each hour of television he watches, that minimum number of hours increases by 0.25. This week, the inequality for his hours of chores is h ≥ 5.5. What does this indicate about his television watching?

Ⓐ He watched 6 hours of television this week.
Ⓑ He watched 5.5 hours of television this week.
Ⓒ He watched 22 hours of television this week.
Ⓓ He watched no television this week.

11. The initial cost to rent a bike is $5. Each hour the bike is rented costs $2. Liz is going to rent a bike and can spend at most $17. Write and solve an inequality to find how long she can rent the bike.

12. Some friends put their money together to buy frozen treats from an ice cream truck. Two people got popsicles. Two people got ice cream cones. Each ice cream cone costs twice as much as popsicle. Two people ordered double scoops. Each double scoop cost three times as much as a popsicle. The total cost for their treats was less than $24.12. What are two possible prices for a popsicle? Write and solve the inequality.

Ⓐ $2.03
Ⓑ $2.02
Ⓒ $2.01
Ⓓ $2.00
Ⓔ $1.95

13. **A garden center sells 21 trays of red flowers, 12 trays of yellow flowers, and 16 trays of pink flowers every day. The gardener wants to know how many days, d, it will take to sell more than 200 trays of flowers. Which inequality models the situation.**

Ⓐ 21d + 12d + 16d <200
Ⓑ 21d + 12d + 16d >200
Ⓒ 200 ≤ 21d + 16d + 12d
Ⓓ 200 ≥ 21d + 16d + 12d

CHAPTER 3 → Lesson 4: Quantitative Relationships

1. **Logan loves candy! He goes to the store and sees that the bulk candy is $0.79 a pound. Logan wants to buy p pounds of candy and needs to know how much money (m) he needs. Which equation would be used to figure out how much money Logan needs?**

   Ⓐ $m = 0.79 \div p$
   Ⓑ $m = 0.79(p)$
   Ⓒ $0.79 = m(p)$
   Ⓓ $m = 0.79 + p$

2. **Logan loves candy! He goes to the store and sees that the bulk candy is $0.84 a pound. Logan wants to buy 3 pounds of candy. Using the equation m = 0.84(p), figure out how much money (m) Logan needs.**

   Ⓐ $1.68
   Ⓑ $2.52
   Ⓒ $2.25
   Ⓓ $2.54

3. **Norman is going on a road trip. He has to purchase gas so that he can make it to his first destination. Gas is $3.55 a gallon. Norman gets g gallons. Which equation would Norman use to figure out how much money (t) it cost to get the gas?**

   Ⓐ $t = g(3.55)$
   Ⓑ $t = g \div 3.55$
   Ⓒ $t = 3.55 \div g$
   Ⓓ $t = g + 3.55$

4. **Norman is going on a road trip. He has to purchase gas so that he can make it to his first destination. Gas is $3.58 a gallon. Norman needs to get 13 gallons. Using the expression t = g(3.58), figure out how much Norman will spend on gas.**

   Ⓐ $46.15
   Ⓑ $39.54
   Ⓒ $46.45
   Ⓓ $46.54

5. **Penny planned a picnic for her whole family. It has been very hot outside, so she needs a lot of lemonade to make sure no one is thirsty. There are 60 ounces in each bottle. Penny purchased b bottles of lemonade. She wants to figure out the total number of ounces (o) of lemonade she has. Which equation should she use?**

   Ⓐ $b = 60(o)$
   Ⓑ $60 = o \times b$
   Ⓒ $60(b) = o$
   Ⓓ $60 = b \div o$

6. **Ethan is playing basketball in a tournament. Each game lasts 24 minutes. Ethan has 5 games to play. Which general equation could he use to help him figure out the total number of minutes that he played? Let t = the total time, g = the number of games, and m = the time per game.**

   Ⓐ t = g + m
   Ⓑ t = g(m)
   Ⓒ t = g − m
   Ⓓ t = g ÷ m

7. **The Spencers built a new house. They want to plant trees around their house. They want to plant 8 trees in the front yard and 17 in the backyard. The trees that the Spencer's want to plant cost $46 each. Could they use the equation t = c(n) where t is the total cost, c is the cost per tree, and n is the number of trees purchased, to figure out the cost to purchase trees for both the front and the back yards?**

   Ⓐ No, because the variables represent only two specific numbers that will never change.
   Ⓑ Yes, because the variables represent only two specific numbers that will never change.
   Ⓒ No, because the variables can be filled in with any number.
   Ⓓ Yes, because the variables can be filled in with any number.

8. **Liz is a florist. She is putting together b bouquets for a party. Each bouquet is going to have f sunflowers in it. The sunflowers cost $3 each. Which equation can Liz use to figure out the total cost (c) of the sunflowers in the bouquets?**

   Ⓐ c = 3(bf)
   Ⓑ c = bf ÷ 3
   Ⓒ c = 3b + f
   Ⓓ c = 3(b + f)

9. **Liz is a florist. She is putting together 5 bouquets for a party. Each bouquet is going to have 6 sunflowers in it. The sunflowers cost $3 each. Using the equation c = 3(bf), figure out how much Liz will charge for the bouquets.**

   Ⓐ $90
   Ⓑ $30
   Ⓒ $18
   Ⓓ $80

10. Hen B lays 4 times as many eggs as Hen A. (Let a = Number of eggs Hen A lays, b = Number of eggs Hen B lays)

Hen A	Hen B
2	8
4	16
7	28
11	44

Which equation represents this scenario?

Ⓐ $b = 4 \div a$
Ⓑ $b = 4 + a$
Ⓒ $b = 4a$
Ⓓ $b = 4a - 4$

11. Which of the following represent balloons that have a cost of $2.50 for a quantity of 10? Choose all that apply.

Ⓐ 20 balloons cost $1.25
Ⓑ 100 balloons $25.00
Ⓒ 30 balloons cost $7.50
Ⓓ 80 balloons cost $8.00

12. Sodas cost $1.25 at the vending machine. Complete the table to show the quantity and total cost of sodas purchased.

Day	Money Spent on Sodas	Sodas Purchased	Price per Soda
Monday		24	$1.25
Wednesday	$57.50		$1.25
Friday	$41.25		$1.25

End of Algebra

Chapter 3: Algebra

Lesson 1: Applying Properties to Rational Expressions

Question No.	Answer	Detailed Explanations
1	A	Remember: "Younger than" is a key phrase that will indicate subtraction. If A is the age of Ruby's brother, and she is 2 years younger than her brother. The correct expression is A - 2.
2	C	8n - (3n-6) = 8n - 3n + 6 = 5n + 6. 5n + 6 is the correct answer.
3	A	6t + (3t - 5) = Remove parentheses: 6t + (3t – 5) = 6t + 3t - 5. Now combine like terms: 6t + 3t – 5 = 9t – 5. 9t - 5 is the correct answer.
4	B	7x - 14x + 21x - 2. Here, I will combine like terms first. 7x - 14x + 21x = 14x. Now we have 14x – 2, which factors into 2(7x - 1). 2(7x - 1) is the correct answer.

5	A	Simplify the expression 3(x + 4) - 2
		Step 1: Multiply terms in parentheses by 3
		    3x+12-2
		Step 2: Combine like terms
		    3x+10

6	A	Simplify the expression $(\frac{1}{2})x + (\frac{3}{2})x$ Add the numerators of the fractions and keep the same denominators $(\frac{4}{2})x$ Simplify this fraction we get 2x.
7	A	Simplify the expression 0.25x + 3 - 0.5x + 2 by combining like terms Step 1: 0.25x - 0.5x + 3 + 2 Step 2: -0.25x + 5
8	C	Remember: Addition means more than and 2x represents multiplication, or times. The statement that represents the expression 2x + 4 is four more than two times a number.

Question No.	Answer	Detailed Explanations
9	A	Remember: A half can be represented by the fraction $\frac{1}{2}$, twice indicates multiplication, and less represents subtraction. Considering the numerator (2x-3), means 3 less than twice a number 'x'. Dividing this by half, we get half of three less than twice a number.
10	D	Simplify the expression $\frac{1}{2}$ (2x + 4) - 3 Step 1: Multiply terms in parentheses by $\frac{1}{2}$ (x+2) - 3 Step 2: Combine like terms (x - 1)
11	B	Distributive property is the correct answer. To remove parentheses, we must multiply the quantity outside parentheses to each term inside parentheses.
12	D	Perimeter = 2 • length + 2 • width 2(4x + 9) + 2(3x - 5) = 8x + 18 + 6x - 10 = 14x +8 ft. 14x + 8 is the correct answer.
13	C	5280 ft ÷ (5x + 3) ft represents the number of laps in 1 mile. 5280 / (5x + 3) is the correct answer.
14	B	5x + 10y + 0(z) = 5x + 10y When 0 is multiplied by any quantity, it produces 0; so 0(z) = 0. Then 5x + 10y + 0 = 5x + 10y 5x + 10y is the correct answer.
15	D	The associative property of addition states that the sum of 3 or more numbers is the same regardless how the numbers are grouped.
16	30x - 18	Distribute the six by multiplying 6 and 5x. Then multiply 6 by 3. Subtract the two products.
17	D, E, and F	D, E, and F are the correct options. Find a common factor between 60 and 24. Some common factors are 6 and 4. Divide the two numbers in the expression by the common factor. 60 divided by 6 is 10 and 24 divided by 6 is 4. Also 60 divided by 4 is 15 and 24 divided by 4 is 6. -6 can be a factor if you make sure to change the 10 to negative and the subtraction sign to a +.
18	$88	p=13 so 8(13-2) = 88.

Lesson 2: Modeling Using Equations or Inequalities

Question No.	Answer	Detailed Explanations
1	D	The first step to solving this problem is to figure out the value of t, the number of hours. As you know, 7 am to 3 pm represents 5 hours to 12 pm, then another 3 hours to 3pm, for a total of 8 hours. This gives us W = 3.75(8) = 30, a full tank.
2	D	Let the volume of water = w and volume of milk = m. m/w = 3/8. Rearranging the above equation, m = $\frac{3w}{8}$, m + w = 187 cubic ft. Replacing the value of m by $\frac{3w}{8}$ in the above equation, $\frac{3w}{8}$ + w = 187, $\frac{11w}{8}$= 187, w = $\frac{187*8}{11}$ = 136 cubic ft.
3	D	Remember: The volume of a rectangular box is V = lwh. This lets us set up the following equation: 960 = 12(10)h, 960 = 120h, 8 = h This gives us a height of 8 inches, as indicated.
4	D	Let b = the number of bushels harvested $\frac{24}{12}$ = $\frac{18}{b}$, 24b = 12(18), 24b = 216, b = 9 bushels is the correct answer.
5	C	Let m = Melanie's age; Then 27-m = Roni's Age; m - (27-m) = 3; m - 27 + m = 3; 2m - 27 = 3; 2m = 3 + 27; 2m = 30; m = $\frac{30}{2}$=15 years=Melanie's Age; Roni's Age=27-m=27-15=12 yrs. Alternative Explanation: Among the options, guess in which case the difference in age is 3 years. It is clear that option (C) is the correct answer.
6	B	Solving this problem requires the creation of a simple equation: 2.5(x) = 50 Dividing 50 by 2.5, the answer is 20.
7	B	11 + (8 + 6) = (y + 8) + 6 This equation demonstrates the associative property for addition, which states that when adding three or more quantities, the way that they are grouped does not affect the sum. Therefore, y = 11 because of the associative property of addition.
8	C	p = 2c + 1, p = 2(7) + 1, p = 14 + 1, p = 15 people present 15 is the correct answer.

◀

Question No.	Answer	Detailed Explanations
9	A	Let the volume of salt = x and volume of sugar = y. $\frac{x}{y} = \frac{4}{8}$. Rearranging the above equation, $x = \frac{4y}{8} = \frac{y}{2}$. x + y = 300 cubic ft. Replacing the value of x by $\frac{y}{2}$ in the above equation, $\frac{y}{2}$ + y = 300. $\frac{3y}{2} = 300$. $y = \frac{(300*2)}{3}$. y = 200 cubic ft.
10	A	Apply the rule to each number's position in the sequence: (1) $(8 + 1^2) - 2 \times 1 = 7$, (2) $(8 + 2^2) - 2 \times 2 = 8$, (3) $(8 + 3^2) - 2 \times 3 = 11$, (4) $(8 + 4^2) - 2 \times 4 = 16$, (5) $(8 + 5^2) - 2 \times 5 = 23$. Therefore, sequence 7, 8, 11, 16, 23, ... follows the rule $(8 + t^2) - 2t$.
11	x = 6	The equation is x + 4 = 10. Subtract 4 from both sides and get x = 6.
12	A and D	A and D are correct options. The model shows that 2 x's equals 14 ones. So 2x = 14 or x + x = 14. To solve, divide both sides by 2 and get x = 7.

Lesson 3: Linear Inequality Word Problems

Question No.	Answer	Detailed Explanations
1	C	3000 must be multiplied by y, the years of experience. This amount must then be added to the base salary of $50,000. This sum must be less than or equal to $70,000.
2	B	The weekly increase of 45 is multiplied by the number of weeks. This amount is added to the initial savings of 225. This sum must be greater than or equal to 500.
3	D	The number of minutes of charging time must be multiplied by 2 and divided by 3 to find the percentage of the battery that is charged over that time. This is then added to the initial charge of 20 percent, and the sum must be at least 75.
4	A	Set up an inequality that $74+0.20t \geq 90$. Solving for t, we find that $t \geq 80$.
5	C	Set up an inequality that $100 - 4n \geq 83$. Solving for n, we find that $n \leq 4.25$. Because you can only miss whole numbers of problems, the most he can miss is 4.
6	B	In the inequality, let T be 12, and solve for L. $12 \geq 3 + 0.2L$ ($\geq$ is to be read as greater than or equal to and $\leq$ as less than or equal to). $9 \geq 0.2L$ or $L \leq 9/0.2$ or $L \leq 45$. Therefore Maximum unsupported length for a panel of thickness 12 mm is 45 ft.
7	C	If the number of units sold is 0, the 300 is still present as the minimum amount of the pay. The 25 is multiplied by 0 in the equation and thus negated, but the 300 remains.
8	B	$C=750 - 200h$. If $h = 3.75$, $C = 750 - (200 \times 3.75) = 750 - 750 = 0$ So, if Neil does 3.75 hours of exercise in a week, he will not gain weight. If he does exercise for more than 3.75 hours, he will lose weight.
9	B	Because there are twice as many girls as boys to be admitted, girls must make up $\frac{2}{3}$ of the admitted group. This leaves one third to be boys. One third of 300 is 100.
10	A	His minimum hours of chores increased by 1.5 hours. This must represent 6 hours of TV watching that week .

Question No.	Answer	Detailed Explanations
11	6 hrs	Write the inequality to be: $5 + 2x$ is less than or equal to 17. Subtract 5 from each side and divide by 2 to get 6. $x \leq 6$. So, Liz can rent a bike almost for 6 hours.
12	D and E	D and E are correct options. Let p be the cost of a popsicle. The cost of an ice cream cone is $2p$ and the cost of a double scoop is $3p$. So, $2p + 2(2p) + 2(3p) = 12p$. $12p$ is less than 24.12. Divide each side by 12 and p has to be less than $2.01.
13	B	The flower trays added up have to be more than 200. So, Option (B) is correct.

Lesson 4: Quantitative Relationships

Question No.	Answer	Detailed Explanations
1	B	"m" represents the amount of money that Logan needs. "p" represents the number of pounds that Logan buys. The amount of money Logan needs is found by multiplying the cost of the candy by the number of pounds that Logan buys. $m = 0.79(p)$
2	B	$m = \$0.84(3)$ $m = \$2.52$
3	A	"t" represents the total amount Norman spent. "g" represents the number of gallons that Norman purchased. To find the total amount that Norman spent, multiply the price of the gas by the total number of gallons that Norman purchased. $t = g(3.55)$
4	D	$t = 13$ gallons $(\$3.58)$ $t = 13(3.58)$ $t = 46.54$ $\$46.54$
5	C	"b" represents the number of bottles of lemonade "o" represents the total number of ounces of lemonade To find out the total number of ounces of lemonade that Penny purchased, multiply the number of bottles (b) by the number of ounces in each bottle, 60. $60(b) = o$
6	B	"t" represents the total number of minutes Ethan played "g" represents the number of games "m" represent the number of minutes in each game To find out the total number of minutes Ethan played, multiply the number of games by the number of minutes in each game. $t = g(m)$
7	D	The equation $t = c(n)$ can be used with any numbers. The equation has the number of trees and the cost of one tree as variables. Those quantities, no matter what they are, when multiplied together will always equal the total cost.
8	A	"c" represents the total cost "b" represents the number of bouquets "f" represents the number of sunflowers To find the total cost, multiply the number of bouquets by the number of sunflowers by the price of the sunflowers. $c = 3(bf)$

Question No.	Answer	Detailed Explanations
9	A	$c = 3(bf)$ $c = 3(6)(5)$ $c = 90$ $90
10	C	The number of eggs that Hen B lays depends on the number of eggs that Hen A lays. Hen B lays 4 times more than Hen A. That is represented as 4a $b = 4a$

11	B & C	B. 100 balloons equal 10 packs of 10
		  $2.50 x 10 quantity = $25.00
		
		 C. 30 balloons equal 3 packs of 10
		  $2.50 x 3 = $7.50

Question 12

Day	Money Spent on Sodas	Sodas Purchased	Price per Soda
Monday	$30.00	24	$1.25
Wednesday	$57.50	46	$1.25
Friday	$41.25	33	$1.25

$m = $30.00
24 x $1.25 = $30
$w = 46$ sodas
$57.50 / $1.25 = 46
$f = 33$ sodas
$41.25/ $1.25 = 33

Chapter 4
Geometry & Measurement

Lesson 1: Circles

1. **A circle is divided into 4 equal sections. What is the measure of each of the angles formed at the center of the circle?**

    Ⓐ 25°
    Ⓑ 180°
    Ⓒ 90°
    Ⓓ 360°

2. **What is the area of a circle with diameter 8 cm? Round your answer to the nearest tenth. Use π = 3.14.**

    Ⓐ 201.1 cm²
    Ⓑ 201.0 cm²
    Ⓒ 50.2 cm²
    Ⓓ 25.1 cm²

3. **What is the radius of a circle with a circumference of 125 cm? Round your answer to the near-est whole number. Use π = 3.14.**

    Ⓐ 24 cm
    Ⓑ 10 cm
    Ⓒ 20 cm
    Ⓓ 19 cm

4. **What is the circumference of a circle with radius 0.5 feet? Round your answer to the nearest tenth. Use π = 3.14.**

    Ⓐ 3.1 ft
    Ⓑ 3.2 ft
    Ⓒ 0.8 ft
    Ⓓ 0.7 ft

5. **Which of the following could constitute the area of a circle?**

    Ⓐ 50 units
    Ⓑ 1 square unit
    Ⓒ 1.5 cubic units
    Ⓓ One half of a unit

6. If two radii form a 30 degree angle at the center of a circle with radius 10 cm, what is the area between them? Round your answer to the nearest tenth. Use π = 3.14.

   Hint: A circle "sweeps out" 360 degrees.

   Ⓐ 26.2 square centimeters
   Ⓑ 26.1 square centimeters
   Ⓒ 314.1 square centimeters
   Ⓓ 314.2 square centimeters

7. What is the area of a circle with radius 2.8 cm? Round your answer to the nearest tenth. Use π = 3.14.

   Ⓐ 24.7 cm²
   Ⓑ 24.6 cm²
   Ⓒ 17.6 cm²
   Ⓓ 8.8 cm²

8. What is the radius of a circle with area 50 square cm? Round your answer to the nearest whole number. Use π = 3.14.

   Ⓐ 4 cm
   Ⓑ 3 cm
   Ⓒ 8 cm
   Ⓓ 7 cm

9. What is the length of a semi-circle (curved part only) with radius 8 in? Round your answer to the nearest tenth. Use π = 3.14.

   Hint: A semi-circle is half a circle.

   Ⓐ 201.0 in
   Ⓑ 50.2 in
   Ⓒ 100.5 in
   Ⓓ 25.1 in

10. What is the diameter of a circle with an area of 50.24 m²? Use π = 3.14.

   Ⓐ 4 m
   Ⓑ 6 m
   Ⓒ 8 m
   Ⓓ 16 m

11. **A circular frisbee has a circumference of 15.2 inches. What is the area of this frisbee?**

    Use 3.14 for pi. Round to the nearest whole number if needed.

12. **Match the description with the correct radius that corresponds with the area or circumference. Use π = 3.14.**

	Radius = 4in.	Radius = 5in.
The circumference of a circle is 31.4 in. Use 3.14 for π.	○	○
The area of a circle is 78.5 in². Use 3.14 for π.	○	○
The area of a circle is 50.24 in². Use 3.14 for π.	○	○
The circumference of a circle is 25.12 in. Use 3.14 for π.	○	○

13. **What is the area of a circle with a diameter of 24in.?**

    Use 3.14 for pi. Round to the nearest whole number if needed.

CHAPTER 4 → Lesson 2: Finding Area, Volume, & Surface Area

1. **Find the area of the rectangle shown below.**

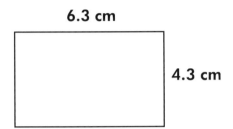

6.3 cm

4.3 cm

Ⓐ 10.5 square centimeters
Ⓑ 24 square centimeters
Ⓒ 27.09 square centimeters
Ⓓ 21 square centimeters

2. **What is the volume of a cube whose sides measure 8 inches?**

Ⓐ 24 in³
Ⓑ 64 in³
Ⓒ 128 in³
Ⓓ 512 in³

3. **Calculate the area of the following polygon.**

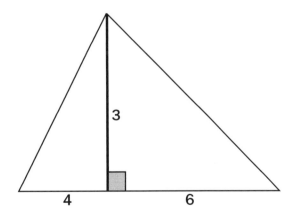

3

4   6

Ⓐ 15 square units
Ⓑ 30 square units
Ⓒ 36 square units
Ⓓ 18 square units

4. **Calculate the area of the following polygon.**

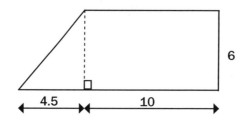

    Ⓐ 60 square units
    Ⓑ 73.5 square units
    Ⓒ 13.5 square units
    Ⓓ 24 square units

5. **What is the volume of the following triangular prism?**

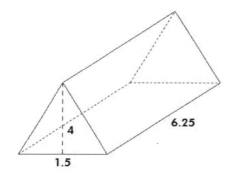

    Ⓐ 11.75 cubic units
    Ⓑ 20 cubic units
    Ⓒ 37.5 cubic units
    Ⓓ 18.75 cubic units

6. **What is the volume of a prism with the following base and a height of 2.75?**

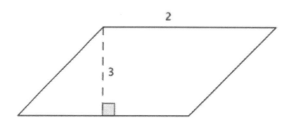

    Ⓐ 8.25 cubic units
    Ⓑ 13.75 cubic units
    Ⓒ 16.5 cubic units
    Ⓓ 8.75 cubic units

7. **What is the surface area of a cube with sides of length 2?**

   Ⓐ 16 square units
   Ⓑ 8 square units
   Ⓒ 24 square units
   Ⓓ 18 square units

8. **What is the surface area of a rectangular prism with dimensions 2, $\dfrac{1}{2}$, and $\dfrac{1}{4}$?**

   Ⓐ 2

   Ⓑ $\dfrac{13}{4}$

   Ⓒ $\dfrac{9}{4}$

   Ⓓ $\dfrac{3}{2}$

9. **Find the area of the shape below. (Round to the nearest tenth). Use pi = 3.14.**

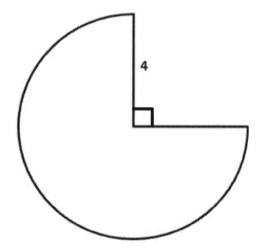

   Ⓐ 37.7 square units
   Ⓑ 50.2 square units
   Ⓒ 18.8 square units
   Ⓓ 35.2 square units

10. John has a container with a volume of 170 cubic feet filled with sand. He wants to transfer his sand into the new container shown below so he can store more sand. After he transfers the sand, how much more sand is remaining in the old container?

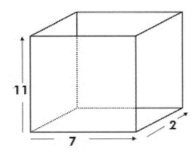

Ⓐ 16 cubic feet of sand
Ⓑ 26 cubic feet of sand
Ⓒ 150 cubic feet of sand
Ⓓ 324 cubic feet of sand

11. What is the area of the shape? Write your answer in the box given below.

56 in.

123 in.

12. Match the question with the correct type of measurement.

	Area	Surface Area	Volume
How much liquid does this bucket hold?	○	○	○
How much wrapping paper do I need to cover or wrap a box?	○	○	○
How much carpet do I need to cover the floor of this room?	○	○	○

13. Find the volume of a right pyramid that has a height of 14 in. and a base area of 25 in². Round your answer to the nearest whole number and write the answer in the box.

14. Match the formula to the correct term.

Area of a triangle	
Volume of a cube	
Area of rectangle	

$$S^3 \qquad l.w \qquad \tfrac{1}{2}.l.bh$$

15. What is the volume of a rectangular prism with dimensions 6, $\dfrac{1}{2}$, $\dfrac{1}{4}$ feet?

CHAPTER 4 → Lesson 3: Cross Sections of 3-D Figures

1. The horizontal cross section of a square pyramid is a _____.

   Ⓐ Square
   Ⓑ Circle
   Ⓒ Trapezoid
   Ⓓ Triangle

2. In order for a three-dimensional shape to be classified as a "prism," its horizontal cross-sections must be _____.

   Ⓐ congruent polygons
   Ⓑ non-congruent polygons
   Ⓒ circles
   Ⓓ equilateral triangles

3. Which of the following nets is NOT the net of a cube?

   Ⓐ

   Ⓑ

   Ⓒ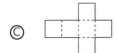

   Ⓓ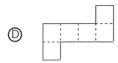

4. The horizontal cross section of a pentagonal prism is a _____.

   Ⓐ Triangle
   Ⓑ Pentagon
   Ⓒ Trapezoid
   Ⓓ Square

5. The vertical cross section of a cube is a _____ .

Ⓐ Square
Ⓑ Triangle
Ⓒ Circle
Ⓓ Trapezoid

6. Which of the following shapes represents the lateral faces of a pyramid?

Ⓐ Triangle
Ⓑ Rectangle
Ⓒ Trapezoid
Ⓓ Square

7. How many faces does a cube have?

Ⓐ 4
Ⓑ 2
Ⓒ 6
Ⓓ 8

8. Which of the following sets of information would not allow you to draw a unique triangle?

Ⓐ The length of the three sides
Ⓑ The length of two of the sides and the angle between them
Ⓒ The size of two angles and one of the sides
Ⓓ The length of two sides and an angle not between them

9. Which statement is not true?

Ⓐ The right cross section of a square pyramid is a trapezoid.
Ⓑ The vertical cross section of a square pyramid is a triangle.
Ⓒ The base of a square pyramid is a triangle.
Ⓓ The horizontal cross section of a square pyramid is a square.

10. Samuel and Christina made a model colonial-style house for their history class. The figures below show the top, side, and front views of the 2 three-dimensional figures they used to make their house.

Based on the figures, which two 3D figures did Samuel and Christina use?

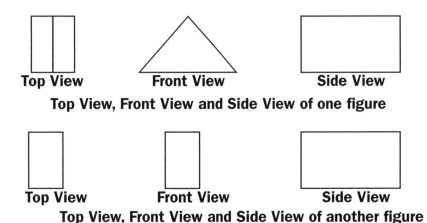

Top View, Front View and Side View of one figure

Top View, Front View and Side View of another figure

Ⓐ Triangular prism and cube
Ⓑ Rectangular prism and triangular prism
Ⓒ Triangular pyramid and triangular prism
Ⓓ Cube and triangular pyramid

11. What are the dimensions of a horizontal cross section of the rectangular prism? Remember that the figure is not drawn to scale.

Select all the correct answers.

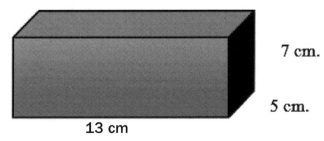

7 cm.

5 cm.

13 cm

Ⓐ 7 cm x 5 cm
Ⓑ 5 cm x 7 cm
Ⓒ 13 cm x 5 cm
Ⓓ 7 cm x 13 cm
Ⓔ 13 cm x 7 cm
Ⓕ 5 cm x 13 cm

12. **What are the dimensions of a vertical cross section (Take the vertical plane cutting across the prism to be parallel to the right side face of the prism) of the rectangular prism? Remember that the figure is not drawn to scale.**

Write your answer in the box given below.

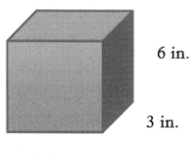

6 in.

3 in.

5 in.

13. **Match the cross section with the description of the shape of the cross section.**

	Rectangle	Square
Horizontal cross section of a rectangular prism	○	○
Vertical cross section of a rectangular prism	○	○
Horizontal cross section of a square pyramid	○	○
Vertical cross section of a cube	○	○

CHAPTER 4 → Lesson 4: Angles

1. **Find x.**

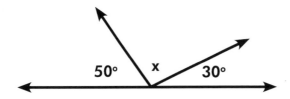

Ⓐ 40°
Ⓑ 60°
Ⓒ 80°
Ⓓ 100°

2. **Find the measures of the missing angles in the figure below.**

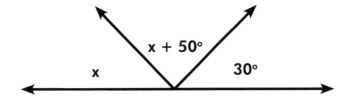

Ⓐ 30° and 60°
Ⓑ 60° and 90°
Ⓒ 50° and 100°
Ⓓ 60° and 120°

3. **The sum of the measures of angles a and b 155 degrees. What is the measure of angle b?**

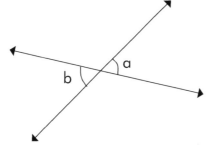

Ⓐ 155 degrees
Ⓑ 77.5 degrees
Ⓒ 35 degrees
Ⓓ 210.5 degrees

4. **What is true about every pair of vertical angles?**

   Ⓐ They are supplementary.
   Ⓑ They are complementary.
   Ⓒ They are equal in measure.
   Ⓓ They total 360 degrees.

5. **If the sum of the measures of two angles is 180 degrees, they are called —**

   Ⓐ supplementary angles
   Ⓑ complementary angles
   Ⓒ vertical angles
   Ⓓ equivalent angles

6. **If angle a measures 30 degrees, what is the measure of angle b?**

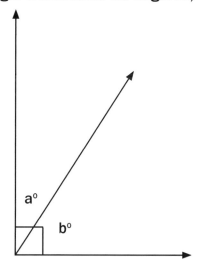

   Ⓐ 60 degrees
   Ⓑ 30 degrees
   Ⓒ 150 degrees
   Ⓓ 20 degrees

7. **If the measure of the first of two complementary angles is 68 degrees, what is the measure of the second angle?**

   Ⓐ 68 degrees
   Ⓑ 22 degrees
   Ⓒ 44 degrees
   Ⓓ 34 degrees

8. If the sum of the measures of angles a and b is 110 degrees, what is the measure of angle c?

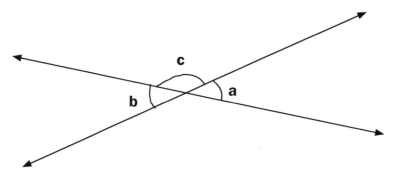

   Ⓐ 125 degrees
   Ⓑ 55 degrees
   Ⓒ 70 degrees
   Ⓓ 180 degrees

9. If two angles are both supplementary and equal in measure, they must be

   Ⓐ vertical angles
   Ⓑ right angles
   Ⓒ adjacent angles
   Ⓓ obtuse angles

10. If the sum of the measures of angles a and b is 240 degrees, what is the measure of angle c?

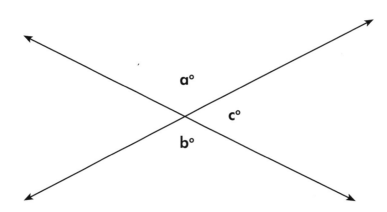

   Ⓐ 60 degrees
   Ⓑ 30 degrees
   Ⓒ 160 degrees
   Ⓓ 150 degrees

11. What is the measure of angle ∠A? Type the answer in the box.

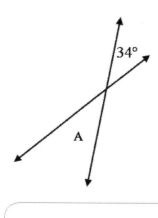

12. What is the measure of angle ∠A? Type the answer in the box.

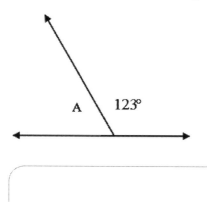

13. What is the measure of angle ∠A? Type the answer in the box.

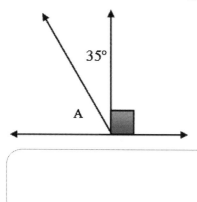

CHAPTER 4 → Lesson 5: Scale Models

1. **Triangle ABC and triangle PQR are similar. Find the value of x.**

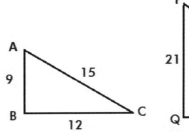 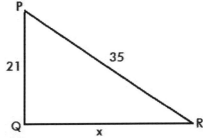

- Ⓐ 23
- Ⓑ 25
- Ⓒ 26
- Ⓓ 28

2. **If the sides of two similar figures have a similarity ratio of $\dfrac{3}{2}$ what is the ratio of their areas?**

- Ⓐ $\dfrac{9}{4}$

- Ⓑ $\dfrac{3}{2}$

- Ⓒ $\dfrac{1}{3}$

- Ⓓ $\dfrac{3}{1}$

3. What is the similarity ratio between the following two similar figures?

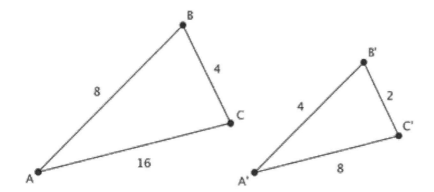

Ⓐ $\dfrac{2}{1}$

Ⓑ $\dfrac{1}{4}$

Ⓒ $\dfrac{2}{3}$

Ⓓ $\dfrac{3}{2}$

4. A map is designed with a scale of 1 inch for every 5 miles. If the distance between two towns is 3 inches on the map, how far apart are they?

Ⓐ 15 miles
Ⓑ 3 miles
Ⓒ 5 miles
Ⓓ 1.5 miles

5. If the angles of one of two similar triangles are 30, 60, and 90 degrees, what are the angles for the other triangle?

Ⓐ 60, 120, 180
Ⓑ 45, 45, 90
Ⓒ 30, 60, 90
Ⓓ There is not enough information to determine.

6. The ratio of similarity between two figures is $\frac{4}{3}$.

   If one side of the larger figure is 12 cm long, what is the length of the corresponding side in the smaller figure?

   Ⓐ 9 cm
   Ⓑ 16 cm
   Ⓒ 3 cm

   Ⓓ $\frac{4}{3}$ cm

7. If the sides of two similar figures have a similarity ratio of $\frac{5}{3}$ what is the ratio of their perimeters?

   Ⓐ $\frac{25}{9}$

   Ⓑ $\frac{5}{6}$

   Ⓒ $\frac{5}{3}$

   Ⓓ $\frac{3}{5}$

8. Triangle ABC and triangle DEF are similar. Find the value of x.

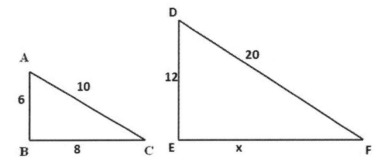

   Ⓐ 16
   Ⓑ 18
   Ⓒ 14
   Ⓓ 12

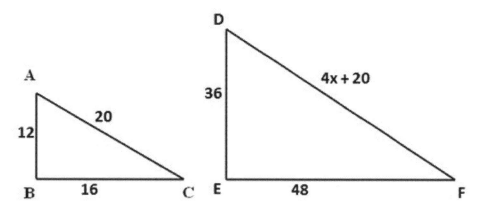

9. Triangle ABC and triangle DEF are similar. Find the value of x.

Ⓐ 5
Ⓑ 20
Ⓒ 10
Ⓓ 15

10. Triangle ABC and triangle DEF are similar. Find the value of x.

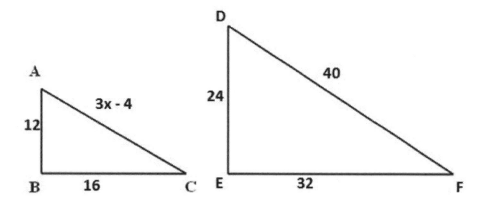

Ⓐ 5
Ⓑ 10
Ⓒ 8
Ⓓ 6

11. On a map, 1 inch equals 15 miles. Two cities are 6 inches apart on the map. What is the actual distance between the cities?

    Write your answer in the box given below

    [   ]

12. On a map, 1.5 inches equals 150 miles. The distance that a man travels is 4 inches on the map. Represent the scale as two different ratios.

    Note: Select two ratios that represent the scale of the map.

    (A) $\dfrac{150 \text{ miles}}{1.5 \text{ inches}}$

    (B) $\dfrac{1.5 \text{ inches}}{150 \text{ miles}}$

    (C) $\dfrac{4 \text{ inches}}{150 \text{ miles}}$

    (D) $\dfrac{150 \text{ miles}}{4 \text{ inches}}$

13. Mark the map with the correct scale for the situation described.

	Map 1 Scale is 1 in : 15 miles	Map 2 Scale is 2 in : 25 miles
Frank traveled an actual distance of 45 miles. On the map, he traveled 3 inches.	○	○
Two cities are 135 miles apart. On the map, they are 9 inches apart.	○	○
It is 137.5 miles from City 1 to City 2. On the map, City 1 and City 2 are 11 inches apart.	○	○
The map shows that the school and library are 5 inches apart. Therefore they are 62.5 miles apart.	○	○

End of Geometry & Measurement

Chapter 4
Geometry & Measurement

Lesson 1: Circles

Question No.	Answer	Detailed Explanations
1	C	The correct answer is 90°. Selecting 25° results from incorrectly applying the qualities of a circle graph to the circle (a whole circle represents 100%, so four equal parts equal 25% each). Choosing 180° is a result of measuring the angle of the line formed by dividing the circle into two equal parts (a straight line measures 180°). A circle measures 360°, so the measure of each angle formed at the center would be less than 360°. By dividing it into 4 equal pieces, each angle will be $360 \div 4 = 90°$.
2	C	The correct answer is 50.2 cm². To find the area of a circle, apply the formula $A = \pi r^2$. Since the problem gives the diameter of the circle, the first step is to find the radius by dividing the diameter by 2. $8 \div 2 = 4$ cm. Next, plug in the numbers into the formula: (1) $\pi 4^2 =$ (2) $3.14 \times 4^2 =$ (3) $3.14 \times 16 =$ (4) 50.24 cm² (5) 50.2 cm² (rounded to the nearest tenth). Common errors made when applying the area formula to circles are multiplying the radius by 2 instead of by itself (which would result in 25.1 cm²) or using the diameter of the circle to find the area (resulting in 201.0 cm²).
3	C	The correct answer is 20 cm. To find the radius when given the circumference of the circle, use $\frac{C}{2\pi}$, where C equals circumference. Insert the numbers from the problem, and solve: (1) $r = \frac{C}{2\pi}$ (2) $125 \div (2 \times 3.14) =$ (3) $125 \div 6.28 =$ (4) 19.90 cm = (5) 20 cm (rounded to the nearest whole number). Finding a radius of 19 cm results from rounding down instead of rounding up. An answer of 10 cm results from dividing the radius by 2. Choosing 24 cm is based on adding $2 + \pi$ and dividing 125 by the sum.

Question No.	Answer	Detailed Explanations
4	A	The answer is 3.1 ft. Finding 3.2 ft results from rounding up instead of rounding down. A result of 0.8 ft comes from applying the formula for the area of a circle. Selecting 0.7 ft also results from using the area formula but also includes a rounding error. To apply the formula for the circumference of a circle, $2\pi r$, plug in the given values, and solve. (1) $2\pi r$ (2) $2 \times 3.14 \times 0.5$ (3) $6.28 \times 0.5 =$ (4) 3.14 ft $=$ (5) 3.1 ft (rounded to the nearest tenth).
5	B	The answer is 1 square unit. Area is defined as the number of square units that cover a specific space. Since the question calls for the area of a circle, the best answer is 1 square unit.
6	A	The answer is 26.2 square centimeters. First, find the area of the entire circle by applying the area formula —> $A = \pi r^2$. (1) $A = 3.14 \times 10^2 = 3.14 \times 100 = 314$ cm^2. Next, calculate the number of square centimeters per degree by dividing the area of the circle by the number of degrees in the circle. (2) $314 \div 360 = 0.87$ cm^2 per degree. Multiply the quotient by 30 and round to the nearest tenth. (3) $30 \times 0.87 = 26.16$ square centimeters (4) 26.2 cm^2 (rounded to the nearest tenth).
7	B	The correct answer is 24.6 cm^2. Area of a circle is given by $A = \pi r^2$, plug in the given values, and solve. $A = 3.14 \times 2.8^2 = 3.14 \times 7.84 = 24.61$ cm^2. Thus, area of the circle is 24.6 cm^2 (rounded to the nearest tenth).
8	A	The answer is 4 cm. To apply the formula for finding radius when given area of a circle, which is the square root of A/π, plug in the given values and solve. r = square root of A/π = square root of $(50 \div 3.14)$ = square root (15.92) = 3.99 cm. Thus, r = 4 cm (rounded to the nearest whole number).
9	D	The answer is 25.1 in. To apply the formula for circumference of a circle, $2\pi r$, plug in the given values and solve. Then, divide the circumference by 2 to find the circumference of the semi-circle. Circumference of Circle = $2\pi r$ C = $2 \times 3.14 \times 8$ = $6.28 \times 8 = 50.24$ Circumference of semicircle = $50.24 \div 2$ = 25.12 in = 25.1 in (rounded to the nearest tenth) Alternate Method : Perimeter of a semi-circle = $\pi r = 3.14 \times 8 = 25.12 = 25.1$ in (rounded to the nearest tenth).

Question No.	Answer	Detailed Explanations
10	C	An answer of 4 m results from calculating the radius by dividing 50.24 ÷ 3.14 and then finding the square root of the quotient, 16. Selecting 6 m results from following the preceding steps, but then adding 4 + 2. Choosing 16 m is the result of dividing 50.24 ÷ 3.14. In order to find the diameter, apply the formula for diameter after finding the radius. The formula is d = 2r (diameter = 2 × radius). As mentioned, the radius is 4 m. Therefore, 4 × 2 = 8 m.
11	18	The formula for area is πr^2 (where r is radius). The formula for circumference is πd (where d is the diameter). If the circumference is 15.2 then if you divide by π, you can find the d. Then d=4.8407. Then the radius r=2.42. Therefore the area is $\pi(2.42)^2$. Therefore the area is 18 in^2 (rounded to the nearest whole number).

12

	Radius = 4 in.	Radius = 5 in.
The circumference of a circle is 31.4 in. Use 3.14 for π.		●
The area of a circle is 78.5 in². Use 3.14 for π.		●
The area of a circle is 50.24 in². Use 3.14 for π.	●	
The circumference of a circle is 25.12 in. Use 3.14 for π.	●	

Use the formulas for area= πr^2 and circumference= $2\pi r$.

(1) radius = circumference / 2π. radius=31.4 / (2 x 3.14)=31.4/6.28=5 in.
(2) radius = sq. root (area / π). radius=sq. root (78.5/3.14)=sq. root (25)=5 in.
(3) radius = sq. root (area / π). radius=sq. root (50.24/3.14)=sq. root (16)=4in.
(4) radius = circumference / 2π. radius=25.12 / (2 x 3.14)=25.12/6.28=4 in.

| 13 | 452 | The formula for area is πr^2 (where r is radius). If d=24. Then the radius r=12. Therefore the area is $\pi(12)^2$. Therefore the area is 452 in^2 (rounded to the nearest whole number). |

Lesson 2: Finding Area, Volume, & Surface Area

Question No.	Answer	Detailed Explanations
1	C	The answer is 27.09 square centimeters. To calculate the area of a rectangle, multiply the length and width. Multiplying $6.3 \times 4.3 = 27.09$ square centimeters.
2	D	The answer is 512 in³. The formula for the volume for a cube is $V = s^3$, where s is the length of one side. Multiplying $8 \times 8 \times 8 = 512$ in³.
3	A	The correct answer is 15 square units. To find the correct answer, apply the formula for the area of a triangle, $A = \frac{1}{2}bh$. First, calculate the base by adding $6 + 4 = 10$. Next, multiply the base times the height: $10 \times 3 = 30$. Then, divide the product by 2: $30 \div 2 = 15$ square units.
4	B	The answer is 73.5 square units. This is a compound figure comprised of a triangle and a rectangle. Identify the length and width of rectangle. Apply the formula for area of a rectangle, $A = bh$ or $A = lw$, and the formula for the area of a triangle, $A = \frac{1}{2}bh$. Then, add the two products together. The sum is the total area of the figure. (1) area of rectangle: $6 \times 10 = 60$ square units (2) area of triangle : $(\frac{1}{2}) \times 4.5 \times 6 = 13.5$ square units (3) $60 + 13.5 = 73.5$ square units
5	D	The answer is 18.75 cubic units. The formula for volume is $V = BH$, where B = the area of the base, and H = height. Since the base is a triangle, the formula is $V = (\frac{1}{2}bh)(H)$. To solve, plug in the numbers: $V = (\frac{1}{2}bh)(H) = (\frac{1}{2} \times 1.5 \times 4)(6.25) = (3)(6.25) = 18.75$ cubic units.
6	C	The answer is 16.5 cubic units. The formula for volume is $V = BH$, where B = the area of the base, and H = height. Since the base is a parallelogram, the formula is $V = (bh)(H)$. To solve, plug in the numbers: $V = (bh)(H) = (2 \times 3)(2.75) = (6)(2.75) = 16.5$ cubic units.
7	C	A cube has 6 square faces. In this particular cube, each face has an area of $2 \times 2 = 4$ square units. The overall surface area = $6 \times 4 = 24$ square units

Question No.	Answer	Detailed Explanations
8	B	This prism is made up of 6 rectangles. Two of them are 2 by 0.5, two of them are 2 by 0.25, and two of them are 0.5 by 0.25 The surface area = 2(2)(0.5) + 2(2)(0.25) + 2(0.5)(0.25) = 3.25 square units, or $\frac{13}{4}$ square units.
9	A	The answer is 37.7 square units. First, find the area of the entire circle by applying the area formula—πr^2. Next, multiply the area by $\frac{3}{4}$ in order to find the area of the shape. (1) πr^2 (2) $3.14 \times 4^2 =$ (3) $3.14 \times 16 =$ (4) Area = 50.24 cm² (5) $50.24 \times \frac{3}{4} = 37.68 =$ (6) 37.68 = 37.7 cm²
10	A	Volume of the new container = $l \times w \times h$ = 7 x 2 x 11 = 154 cubic feet. Therefore, after transferring the sand, amount of sand remaining in the old container = 170 - 154 = 16 cubic feet.
11	6,888	Use the formula A = L x W

12

	Area	Surface Area	Volume
How much liquid does this bucket hold?			●
How much wrapping paper do I need to cover or wrap a box?		●	
How much carpet do I need to cover the floor of this room?	●		

Volume is the amount of space occupied by a 3D shape. Area is the amount of space occupied by a 2D shape. Surface area is the area of the surfaces of a 3D shape.

| 13 | 117 | The formula for volume of a right pyramid is $\frac{1}{3}$rd of the product of the height and the base area. Therefore $\frac{1}{3}(14 \times 25) = 116.666...$ Rounded to the nearest whole number is 117 in³. |

Question No.	Answer	Detailed Explanations
14		

Area of a triangle ●━━━━━●	$\frac{1}{2}$ x b x h
Volume of a cube ●━━━━━●	S^3
Area of rectangle ●━━━━━●	l x b

(1) Area of a triangle is given by $\frac{1}{2}$ x b x h. Where b is breadth and h is height

(2) Volume of a cube is given by s^3. Where s is the side

(3) Area of a rectangle is given by l x b. Where l is the length and b is the breadth

Question No.	Answer	Detailed Explanations
15	$\frac{3}{4}$	The volume of a rectangular prism is given by $l.b.h$ Hence, volume = $6 \times \frac{1}{2} \times \frac{1}{4} = \frac{3}{4}$ ft^3

Lesson 3: Cross Sections of 3-D Figures

Question No.	Answer	Detailed Explanations
1	A	A square pyramid consists of a square base and 4 triangular sides (for a total of 5 faces); therefore, the cross section of a pyramid would show a square.
2	A	A prism consists of two congruent bases and various congruent faces; therefore, the cross sections of a prism must show congruent polygons.
3	A	In order to form a cube, the nets must fold together to make the shape. Here, the figure would not form a cube when folded together to make a three-dimensional figure.
4	B	Horizontal cross section of a pentagonal prism is a pentagon.
5	A	A cube is a rectangular prism with four lateral faces and two bases (also faces) which are square. Therefore, the cross section of a cube is a square.
6	A	A pyramid consists of 1 polygon base and all triangular lateral faces; therefore, the side of pyramid would show a triangle.
7	C	A cube is a rectangular prism with four lateral faces and two bases (also faces) which are square. Therefore, the total number of faces for a cube is 6.
8	D	The length of two sides and an angle not between them cannot always uniquely determine a triangle. All other three cases (length of three sides, length of two sides and angle between them and the size of two angles and one of the sides) uniquely determine a triangle i.e. only one triangle can be drawn.
9	C	The base of a square pyramid is a square, not a triangle.
10	B	The first 2D figure represents a 3D shape with 6 rectangular faces. The second 2D figure represents a 3D shape with two triangular bases and 3 rectangular faces. Therefore, Samuel and Christina used a rectangular prism and a triangular prism for their house.
11	C and F	The cross section horizontally does not use the height of the figure so the length and width are only used.

Question No.	Answer	Detailed Explanations
12	6 x 3	Two vertical cross sections are possible for a rectangular prism : (1) When the plane of vertical cross section is parallel to the right (and left) face of the prism. Then the resulting rectangle will have the dimensions 6 in. X 3 in. (2) When the plane of vertical cross section is parallel to the front (and back) face of the prism. Then the resulting rectangle will have the dimensions 5 in. X 6 in.

13

	Rectangle	Square
Horizontal cross section of a rectangular prism	●	
Vertical cross section of a rectangular prism	●	
Horizontal cross section of a square pyramid		●
Vertical cross section of a cube		●

The vertical and horizontal cross sections of a rectangular prism are rectangles. The horizontal cross section of a square pyramid is the same shape as the base (square). Also, the horizontal and the vertical cross section of a cube is a square.

Lesson 4: Angles

Question No.	Answer	Detailed Explanations
1	D	The answer is 100°. Reminder: Angles that together form a straight line are called supplementary, meaning they add to 180 degrees. In this case, $50 + x + 30 = 180$ requires an x value of 100 degrees.
2	C	The answer is 50° and 100°. Angles that together form a straight line are supplementary, meaning their measures add to 180 degrees. In this case, $x + (x + 50) + 30 = 180$ can only be satisfied by an x value of 50, resulting in angles of measure 50 and 100 degrees.
3	B	The answer is 77.5 degrees. When two lines intersect, they form vertical angles, which are equal in measure. Angle a and angle b are vertical angles. To find the value of angle b, divide 155 by 2. The quotient is the value of angle b: $155 \div 2 = 77.5$ degrees.
4	C	Vertical angles are congruent angles formed by two intersecting lines. The sum of the angles can be less than or greater than 90 degrees and 180 degrees, respectively, so they are not necessarily complementary or supplementary angles. Also, vertical angles do not form a complete circle, so they do not total 360 degrees.
5	A	Two angles are supplementary if the sum of their measures is equal to 180°. The sum of the measures of the angles cannot be greater than or less than 180°. It must be exactly 180°.
6	A	The answer is 60°. Angles a and b form a right angle, which measures 90°. This means that angles a and b are complementary. To find the measure of angle b, subtract the measure of angle a from 90: $90 - 30 = 60$. Therefore, the measure of angle b is 60°.
7	B	The answer is 22°. Two angles are complementary if the sum of the measures of the angles equals 90°. To find the measure of the second angle, subtract the measure of the first angle from 90: $90 - 68 = 22$. Therefore, the measure of the second angle is 22°.
8	A	The answer is 125 degrees. When two lines intersect to form vertical angles, the opposite angles are equal in measure. Angle a and angle b are vertical angles. To find the value of angle a, divide 110 by 2. The quotient is the value of angle a: $110 \div 2 = 55$ degrees. Two intersecting lines also form adjacent supplementary angles which add up to 180°. Since angle a and angle c are adjacent and supplementary, subtract 55 from 180 to find the measure of angle c: $180 - 55 = 125$.

Question No.	Answer	Detailed Explanations
9	B	Two angles are supplementary if the sum of the measures of the angles equals 180°. Since right angles are 90°, the sum of two right angles is 180°. Therefore, two right angles form supplementary angles.
10	A	The answer is 60 degrees. When two lines intersect to form vertical angles, the opposite angles are equal in measure. Angle a and angle b are vertical angles. To find the value of angle a and angle b, divide 240 by 2: 240 ÷ 2 = 120 degrees. Two intersecting lines also form adjacent supplementary angles, which add up to 180 degrees. Since angle a and angle c are adjacent and supplementary, subtract 120 from 180 to find the measure of angle c: 180 - 120 = 60.
11	34	Vertical angles have the same measure.
12	57	Supplementary angles must be added together to equal 180 degrees.
13	55	Complementary angles must be added together to equal 90 degrees.

Lesson 5: Scale Models

Question No.	Answer	Detailed Explanations
1	D	Remember: In order to solve a similarity question, set up a proportion with corresponding sides, and solve!: $\frac{x}{12} = \frac{35}{15}$ or $\frac{x}{12} = \frac{21}{9}$. You can use cross products (ad = bc) to solve for x. (1) $\frac{x}{12} = \frac{35}{15}$ (2) 15x = (35)(12) (3) 15x = 420 (4) x = 420 ÷ 15 (5) x = 28
2	A	If the similarity ratio is $\frac{3}{2}$, then the ratio of the areas is the square of that ratio: $\frac{3}{2} \times \frac{3}{2} = \frac{9}{4}$.
3	A	In order to solve a similarity problem, set up a proportion with corresponding sides: $\frac{8}{4} = \frac{16}{8}$. Both ratios in simplest form are $\frac{2}{1}$. Therefore, the similarity ratio is $\frac{2}{1}$.
4	A	You can use cross products (ad = bc) to solve for x. (inch)/(miles) = (inch)/(miles) (1) $\frac{1}{5} = \frac{3}{x}$ (2) 1x = (3)(5) (3) x = 15 Therefore, the towns are 15 miles apart.
5	C	Corresponding angles are congruent in similar figures. Thus, the angles will have the same measures in the second figure as in the first.
6	A	larger side / smaller side = $\frac{4}{3}$ or larger side = $(\frac{4}{3})$x smaller side. Therefore, smaller side = larger side ÷ $(\frac{4}{3})$. Therefore, in order to find the length of the smaller side, divide 12 by $\frac{4}{3}$. (1) $\frac{12}{1} \div \frac{4}{3}$ (2) $\frac{12}{1} \times \frac{4}{3}$ (3) $\frac{12 \times 3}{1 \times 4}$ (4) $\frac{36}{4}$ (5) 9. Therefore, the length of the corresponding side is 9 cm.

Question No.	Answer	Detailed Explanations
7	C	Let the sides of the larger triangle be 5a, 5b and 5c units in length. Then the sides of the smaller triangle will be 3a, 3b and 3c. Therefore the ratio of their perimeters will be $\frac{5(a+b+c)}{3(a+b+c)} = \frac{5}{3}$.
8	A	You can use cross products (ad = bc) to solve for x. (1) $\frac{10}{20} = \frac{8}{x}$ (2) 10x = (20)(8) (3) 10x = 160 (divide both sides by 10) (4) x = 16
9	C	You can use cross products (ad = bc) to solve for x. (1) $\frac{16}{48} = \frac{20}{4x + 20}$ (2) (16)(4x + 20) = (48)(20) (3) 64x + 320 = 960 (subtract both sides by 320) (4) 64x = 640 (divide each side by 64) (5) x = 10
10	C	You can use cross products (ad = bc) to solve for x. (1) $\frac{16}{32} = \frac{3x - 4}{40}$ (2) (16)(40) = (32)(3x - 4) (3) 640 = 96x - 128 (add both sides by 128) (4) 768 = 96x (divide each side by 96) (5) 8 = x
11	90	15 miles = 1 inch, so 15 x 6 = 90 miles.
12	A and B	A and B are correct options. The scale of the map is a ratio of miles to inches or inches to miles. It does not matter which order you put them in as long as you use the same order when calculating the proportions for distance for the man. You only use the map's information when calculating the scale of the map.

13

	Map 1 Scale is 1 in: 15 miles	Map 2 Scale is 2 in : 25 miles
Frank traveled an actual distance of 45 miles. On the map, he traveled 3 inches.	●	
Two cities are 135 miles apart. On the map, they are 9 inches apart.	●	
It is 137.5 miles from City 1 to City 2. On the map, City 1 and City 2 are 11 inches apart.		●
The map shows that the school and library are 5 inches apart. Therefore they are 62.5 miles apart.		●

Frank traveled an actual distance of 45 miles. On the map, he traveled 3 inches. Therefore, 1 inch on the map corresponds to 45/3 = 15 miles. So, the scale is 1 in : 15 miles.

Two cities are 135 miles apart. On the map, they are 9 inches apart. Therefore, 1 inch on the map corresponds to 135/9 = 15 miles. So, the scale is 1 in : 15 miles.

City 1 and City 2 are 137.5 miles apart. On the map, they are 11 inches apart. Therefore, 1 inch on the map corresponds to 137.5/11 = 12.5 miles. So, the scale is 1 in : 12.5 miles Or 2 in : 25 miles.

The school and library are 62.5 miles apart. Then The map shows that the school and library are 5 inches apart. Therefore, 1 inch on the map corresponds to 62.5/5 = 12.5 miles. So, the scale is 1 in : 12.5 miles Or 2 in : 25 miles.

Chapter 5: Statistics & Probability

Lesson 1: Mean, Median, and Mean Absolute Deviation

1. **Consider the following dot-plot for Height versus Weight.**

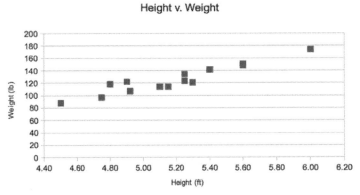

Height v. Weight

    What does this dot-plot indicate about the correlation between height and weight?

    Ⓐ There is no correlation.
    Ⓑ There is a strong negative correlation.
    Ⓒ There is a strong positive correlation.
    Ⓓ There is a weak positive correlation.

2. **The following chart represents the heights of boys on the basketball and soccer teams.**

Basketball	Soccer
5′4″	4′11″
5′2″	4′10″
5′3″	5′9″
5′5″	5′1″
5′5″	5′0″
5′1″	5′1″
5′9″	5′3″
5′3″	5′1″

What inference can be made based on this information?

Ⓐ Soccer players have a higher average skill level than basketball players.
Ⓑ Soccer players have a lower average weight than basketball players.
Ⓒ Basketball players have a higher average height than soccer players.
Ⓓ No inference can be made.

3. Use the table below to answer the question that follows:

Month	Avg Temp.
January	24°F
February	36°F
March	55°F
April	65°F
May	72°F
June	78°F

What is the difference between the mean temperature of the first four months of the year and the mean temperature of the next two months?

Ⓐ 15 degrees
Ⓑ 20 degrees
Ⓒ 25 degrees
Ⓓ 30 degrees

4. Use the table below to answer the question:

Month	Avg Temp.
January	24°F
February	36°F
March	55°F
April	65°F
May	72°F
June	78°F

If the temperature in January was 54°F instead of 24°F, by how much would the mean temperature for the six months increase?

Ⓐ 5°F
Ⓑ 10°F
Ⓒ 30°F
Ⓓ 35°F

5. Use the table to answer the question below:

Team	Wins
Mustangs	14
Spartans	17
North Stars	16
Hornets	9
Stallions	13
Renegades	9
Rangers	5

What is the mode of the wins for all the teams in the above table?

Ⓐ 14
Ⓑ 5
Ⓒ 13
Ⓓ 9

6. Jack scored 7, 9, 2, 6, 15 and 15 points in 6 basketball games. Find the mean, median and mode scores for all the games.

Ⓐ Mean = 7, Median = 9 and Mode = 2
Ⓑ Mean = 9, Median = 7 and Mode = 15
Ⓒ Mean = 9, Median = 8 and Mode = 15
Ⓓ Mean = 7, Median = 2 and Mode = 9

7. Mean absolute deviation is a measure of...

Ⓐ Central Tendency
Ⓑ Variability
Ⓒ Averages
Ⓓ Sample Size

8. Calculate the mean for the following set of data:

$$\left\{ \frac{7}{4}, \frac{3}{4}, \frac{5}{4}, \frac{7}{4}, \frac{3}{4}, \frac{5}{4} \right\}$$

Ⓐ $\dfrac{7}{4}$

Ⓑ $\dfrac{5}{2}$

Ⓒ $\dfrac{5}{4}$

Ⓓ $\dfrac{1}{6}$

9. Another word for mean is...

Ⓐ Average
Ⓑ Middle
Ⓒ Most
Ⓓ Count

10. What is the median for the following set of data?

$$\left\{ \frac{4}{5}, \frac{1}{3}, \frac{1}{3}, \frac{1}{5}, \frac{2}{3} \right\}$$

Ⓐ $\dfrac{1}{3}$

Ⓑ $\dfrac{2}{3}$

Ⓒ $\dfrac{1}{5}$

Ⓓ $\dfrac{1}{4}$

11. **What is the mode of the following set of data?**

{1, 1, 2, 3, 1, 4, 6, 2, 3}

(A) 2
(B) 3.75
(C) 1
(D) 6

12. **John scored 6, 8, 1, 5, 11, 14 and 14 points in 7 lacrosse games. Find the mean, median and mode scores for all the games. (Round to the nearest tenth)**

(A) Mean = 8.4, Median = 8 and Mode = 14
(B) Mean = 8.1, Median = 6, and Mode = 14
(C) Mean = 9.3, Median = 8 and Mode = 15
(D) Mean = 7.5, Median = 2 and Mode = 9

13. **Calculate the mean for the following set of data:**

{ 0.3, 1.2, 2.5, 4.3 }

Round your answer to the nearest tenth.

(A) 2.1
(B) 2.7
(C) 8.3
(D) 2.5

14. **Calculate the median for the following set of data:**

{ 0.3, 1.2, 2.5, 4.3 }

Round your answer to the nearest tenth.

(A) 1.8
(B) 2.1
(C) 1.9
(D) 1.2

15. **What is the mean absolute deviation for the following set of data?**

{2, 5, 7, 1, 2}

(A) 3.4
(B) 2
(C) 2.08
(D) 5

16. Fill in the blanks in the table with the correct value of mean or median. Round the number to the tenth place in case the answer is not a whole number.

	Median	Mean
Data Set : 9, 8, 7, 4, 5		6.6
Data Set : 5, 8, 5, 9, 1, 2	5	
Data Set : 1, 2, 8, 1, 3, 3		3

17. The mean of a data set is 5. The minimum of the data is 2. What is the deviation of the minimum?

Write your answer in the box given below.

CHAPTER 5 → Lesson 2: Mean, Median, and Mode

1. **The following data set represents a score from 1-10 for a customers' experience at a local restaurant.**

   { 1, 1, 2, 1, 3, 4, 7, 8, 1, 3, 4, 2, 1, 3, 7, 2 }

   If a score of 1 means the customer did not have a good experience, and a 10 means the customer had a fantastic experience, what can you infer by looking at the data?

   Ⓐ Overall, customers had a good experience.
   Ⓑ Overall, customers had a bad experience.
   Ⓒ Overall, customers had an "ok" experience.
   Ⓓ Nothing can be inferred from this data.

2. **The manager of a local pizza place has asked you to make suggestions on how to improve his menu. The following bar graph represents the results of a survey asking customers what their favorite food at the restaurant was.**

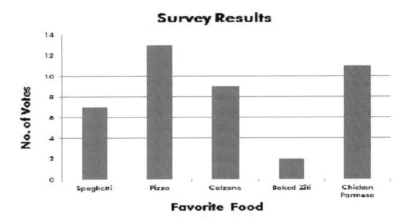

   Based on these survey results, which menu item would you suggest the manager remove from the menu?

   Ⓐ Spaghetti
   Ⓑ Pizza
   Ⓒ Calzone
   Ⓓ Baked Ziti

3. **What are the measures of central tendency?**

   Ⓐ Mean, Median, Mode
   Ⓑ Median, Mode, Mean Absolute Deviation
   Ⓒ Median, Mean Absolute Deviation, Sample Size
   Ⓓ Mean, Median, Range

4. **What is the mean absolute deviation for the following set of data?**

{1, 2, 3, 4}

Ⓐ 1
Ⓑ 2.5
Ⓒ 4
Ⓓ 2

5. **What are the central tendencies of the following data set? (round to the nearest tenth)**

{2, 2, 3, 4, 6, 8, 9, 10, 13, 13, 16, 17}

Ⓐ Mean: 8.6, Median: 8.5, Mode: 2, 13
Ⓑ Mean: 8.5, Median: 8.4, Mode: 2, 3
Ⓒ Mean: 8.5, Median: 8.6, Mode: 2, 13
Ⓓ Mean: 7.6, Median: 7.5, Mode: 10, 13

6. **What are the central tendencies of the following data set? (round to the nearest tenth)**

{11, 11, 12, 13, 15, 17, 18, 20, 23, 23, 26, 27}

Ⓐ Mean: 17.2, Median: 17.5, Mode: none
Ⓑ Mean: 18.5, Median: 16, Mode: 11
Ⓒ Mean: 17.5, Median: 18, Mode: 23
Ⓓ Mean: 18, Median: 17.5, Mode: 11, 23

7. **What are the central tendencies of the following data set? (round to the nearest tenth)**

{31, 31, 32, 33, 35, 37, 38, 41, 54, 54, 57, 58}

Ⓐ Mean: 41.4, Median: 37.5, Mode: 31
Ⓑ Mean: 41.5, Median: 36.5, Mode: none
Ⓒ Mean: 41.6, Median: 38.5, Mode: 31, 54
Ⓓ Mean: 41.8, Median: 37.5, Mode: 31, 54

8. **What are the central tendencies of the following data set? (round to the nearest tenth)**

{41, 41, 42, 43, 45, 47, 48, 51, 64, 64, 64, 67}

Ⓐ Mean: 51.4, Median: 47.5, Mode: 64
Ⓑ Mean: 48.5, Median: 46, Mode: 64
Ⓒ Mean: 47.5, Median: 48, Mode: 64
Ⓓ Mean: 47, Median: 47.5, Mode: 64

9. **What are the central tendencies of the following data set? (round to the nearest tenth)**

   {51, 51, 52, 53, 55, 57, 58, 61, 54, 54, 57, 54}

   Ⓐ Mean: 58.5, Median: 56, Mode: 54
   Ⓑ Mean: 54.8, Median: 54, Mode: 54
   Ⓒ Mean: 57.5, Median: 58, Mode: 54
   Ⓓ Mean: 57.6, Median: 57.5, Mode: 54

10. **What are the central tendencies of the following data set? (round to the nearest tenth)**

    {61, 61, 52, 53, 65, 67, 58, 61, 64, 64, 57, 54}

    Ⓐ Mean: 59.8, Median: 61, Mode: 61
    Ⓑ Mean: 58.5, Median: 66, Mode: 64
    Ⓒ Mean: 57.5, Median: 68, Mode: 64
    Ⓓ Mean: 57.9, Median: 67.5, Mode: 64

11. **What are the central tendencies of the following data set? (round to the nearest tenth)**

    {71, 71, 62, 63, 75, 77, 68, 71, 74, 74, 67, 74}

    Ⓐ Mean: 70.6, Median: 78, Mode: 74
    Ⓑ Mean: 68.5, Median: 76, Mode: 74
    Ⓒ Mean: 70.6, Median: 71, Mode: 71, 74
    Ⓓ Mean: 77, Median: 77.5, Mode: 74

12. **What are the central tendencies of the following data set? (round to the nearest tenth)**

    {61, 71, 52, 63, 65, 77, 58, 71, 64, 74, 57, 74}

    Ⓐ Mean: 65.6, Median: 64.5, Mode: 71, 74
    Ⓑ Mean: 68.5, Median: 66, Mode: none
    Ⓒ Mean: 67.5, Median: 68, Mode: 71
    Ⓓ Mean: 67, Median: 67.5, Mode: 2

13. **What are the central tendencies of the following data set? (round to the nearest tenth)**

    {21, 21, 22, 23, 35, 37, 38, 41, 44, 44, 47, 44}

    Ⓐ Mean: 38.5, Median: 37.5, Mode: 44
    Ⓑ Mean: 34.8, Median: 37.5, Mode: 44
    Ⓒ Mean: 37.5, Median: 38, Mode: none
    Ⓓ Mean: 37, Median: 37.5, Mode: 54

14. Kelli's Ice Cream Shop must have a mean of 110 visitors per day in order to make a profit. The table below shows the number of visitors during one week.

	No of Visitors
Sunday	63
Monday	77
Tuesday	121
Wednesday	96
Thursday	137
Friday	154

Based on the table, how many visitors will need to go to Kelli's Ice Cream shop on Saturday in order for the company to make a profit for the week?

Ⓐ 108
Ⓑ 110
Ⓒ 122
Ⓓ 222

15. Henry made a list of his math scores for one grading period: 98, 87, 93, 89, 90, 85, 88. If Henry adds a 90 to the list to represent his final test for the grading period, which statement is true?

Ⓐ The mode would decrease.
Ⓑ The mean would increase.
Ⓒ The median would increase.
Ⓓ The mean would decrease.

16. Calculate the Mean, Median and Mode for the data set given below.

5, 5, 9, 4, 6, 7

Match the value of the answer to it's correct category.

	Mean	Median	Mode
6	◯	◯	◯
5.5	◯	◯	◯
5	◯	◯	◯

17. Find the interquartile range of the data set.

32.6, 98.5, 16.6, 22.4, 99.8, 72.6, 68.2, 51.8, and 49.3.

CHAPTER 5 → Lesson 3: Sampling a Population

1. **Joe and Mary want to calculate the average height of students in their school. Which of the following groups of students would produce the least amount of bias?**

    Ⓐ Every student in the 8th grade.
    Ⓑ Every student on the school basketball team.
    Ⓒ A randomly selected group of students in the halls.
    Ⓓ Joe & Mary's friends.

2. **Which of the following represents who you should survey in a population?**

    Ⓐ A random, representative group from the population
    Ⓑ Every individual in a population
    Ⓒ Only those in the population that agree with you
    Ⓓ Anyone, including those not in the population

3. **What does increasing the sample size of a survey do for the overall results?**

    Ⓐ Decreases bias in the results
    Ⓑ Increases the mean of the results
    Ⓒ Increases the reliability of the results
    Ⓓ Increasing sample size does not impact the results of a survey

4. **Which of the following is not a valid reason for not surveying everyone in a population?**

    Ⓐ It takes a far longer amount of time to survey everyone.
    Ⓑ Not everyone will be willing to participate in the survey.
    Ⓒ It is hard to determine the exact size of a population necessary to ensure everyone is surveyed.
    Ⓓ Surveying everyone produces unreliable results.

5. **Why is it important to know the sample size of a given survey?**

    Ⓐ It helps determine whether any bias exists.
    Ⓑ It helps determine how reliable the results are.
    Ⓒ It gives a good estimate for the size of the target population.
    Ⓓ It is not important to know the sample size.

6. **John and Maggie want to calculate the average height of students in their school. Which of the following groups of students would most likely produce the most amount of bias?**

    Ⓐ Every student in the 8th grade.
    Ⓑ Every student on the school basketball team.
    Ⓒ A randomly selected group of students in the halls.
    Ⓓ John & Maggie's friends.

7. **Which of the following question types will provide the most useful statistical results?**

    Ⓐ Open-ended questions where the person surveyed can answer in any way they want
    Ⓑ Multiple choice questions offering the person a representative number of choices
    Ⓒ True or false questions

8. **Which of the following does not represent a way of avoiding bias in survey results?**

    Ⓐ Use neutral words in the questions asked
    Ⓑ Ensure a random sample of the population
    Ⓒ Only survey individuals that will answer a certain way
    Ⓓ Tailor the conclusions based on survey results, not previous thoughts

9. **The following data set represents survey results on a scale of 1 to 10.**

    {8, 8, 9, 8, 6, 7, 7, 7, 8, 8, 6}

    Which of the following survey result would you be most surprised with if given by the next person surveyed?

    Ⓐ 6
    Ⓑ 5
    Ⓒ 8
    Ⓓ 7

10. **The following data set represents survey results on a scale of 1 to 10.**

    {6, 6, 7, 6, 8, 7, 7, 7, 6, 6, 8}

    Which of the following survey results would you be most surprised with if given by the next person surveyed?

    Ⓐ 6
    Ⓑ 10
    Ⓒ 8
    Ⓓ 7

11. A company is conducting a survey on their performance with customer service. What method will best avoid receiving biased data?

(A) The company should make the survey anonymous.
(B) The company should take a video survey.
(C) The company should require the survey during the transaction.
(D) The company should just request that everyone fill out a survey.

12. A company is conducting a survey on their performance with customer service. What should the survey look like?

(A) The survey should contain a few simple multiple choice questions with an optional comment section.
(B) The survey should contain simple free response questions with an optional comment section.
(C) The survey should contain only free response questions.
(D) The survey should contain some personal questions.

13. Where should a survey about personal items purchased be conducted?

(A) The receipt should have a link to the survey website.
(B) The survey should be near the register.
(C) The survey should be near the exit doors.
(D) The survey should be in the parking lot.

14. How should online video game surveys be conducted?

(A) This survey should be put at the end of a level or section of the game.
(B) This survey should be put at the beginning of the game.
(C) Thus survey should be put on flyers and distributed at a gaming store.
(D) This survey should be put on the website separate from the game.

15. A large corporation is launching a new product, Hitz, which allows customers to purchase music and store it on the corporation's servers. They want to survey people who listen to a lot of music. Which of the following would give them the best sample?

(A) Conducting a survey inside of a music store
(B) Conducting a survey at the parking lot of a music concert
(C) Conducting an online survey on music social networking websites
(D) Conducting a survey of students in a school marching band.

16. Match the situation with the sampling method used.

	Convenience Sample	Systematic Sample	Simple Random Sample
A person chooses every 5th person on a list of names starting with #1.	○	○	○
A person accepts the first 15 people to respond to a magazine ad.	○	○	○
A person picks names out of a hat.	○	○	○

17. A jar of marbles contains gray and black marbles. You collect the representative sample shown here:

If the jar contains 50 marbles, about how many marbles are gray?

Write your answer in the box given below.

```



```

CHAPTER 5 → Lesson 4: Describing Multiple Samples

1. John comes up with the following methods for generating unbiased samples from shoppers at a mall.

   I. Ask random strangers in the mall

   II. Always go to the mall at the same time of day

   III. Go to different places in the mall

   IV. Don't ask questions the same way to different people

   Which of these techniques represents the best way of generating an unbiased sample?

   Ⓐ I and II
   Ⓑ I and III
   Ⓒ I, II, and III
   Ⓓ All of these

2. These two samples are about students' favorite subjects. What inference can you make concerning the students' favorite subjects?

Student samples	Science	Math	English Language Arts	Total
#1	40	14	30	84
#2	43	17	33	93

   Ⓐ Students prefer Science over the other subjects.
   Ⓑ Students prefer Math over the other subjects.
   Ⓒ Students prefer English language arts over the other subjects.
   Ⓓ Students prefer History over the other subjects.

3. These two samples are about students' favorite types of movies. What inference can you make concerning the students' favorite types of movies?

Student samples	Comedy	Action	Drama	Total
#1	35	45	19	99
#2	38	48	22	108

   Ⓐ Students prefer action movies over the other types.
   Ⓑ Students prefer drama over the other types.
   Ⓒ Students prefer comedy over the other types.
   Ⓓ none

4. These two samples are about students' favorite fruits. What inference can you make concerning the students' favorite fruits?

Student samples	Blueberries	Bananas	Strawberries	Total
#1	33	18	44	95
#2	30	20	40	90

Ⓐ Students prefer strawberries over the other fruits.
Ⓑ Students prefer bananas over the other fruits.
Ⓒ Students prefer blueberries over the other fruits.
Ⓓ none

5. Jane and Matt conducted two surveys about students' favorite sports to play. What inference can you make concerning the students' favorite sports?

Student samples	Soccer	Basketball	Tennis	Total
#1	50	145	26	221
#2	56	150	20	226

Ⓐ Most students like basketball over soccer or tennis.
Ⓑ Most students like soccer over basketball or tennis.
Ⓒ Most students like tennis over soccer or basketball.
Ⓓ Most students like track.

6. Paul and Maggie conducted two surveys about students' favorite seasons. What inference can you make concerning the students' favorite seasons?

Student samples	Summer	Fall	Spring	Total
#1	100	128	244	472
#2	98	129	250	477

Ⓐ Most students prefer the Spring season over Summer and Fall.
Ⓑ Students prefer the Summer season.
Ⓒ Students prefer the Fall season.
Ⓓ Students prefer the Winter season.

7. **John and Mark conducted two surveys about students' favorite pizza toppings. What inference can you make concerning the students' favorite pizza toppings?**

Student samples	pineapple	pepperoni	extra cheese	Total
#1	145	237	118	500
#2	150	230	120	500

Ⓐ Most students like pepperoni.
Ⓑ Most students like extra cheese.
Ⓒ Most students like pineapple.
Ⓓ The students like pepperoni over pineapple or extra cheese.

8. **Jon and Minny conducted two surveys about students' favorite board games. What inference can you make concerning the students' favorite board games?**

Student samples	Game X	Game Y	Game Z	Total
#1	45	5	6	56
#2	50	6	2	58

Ⓐ Most students like Game Z.
Ⓑ Most students like **Game X** over **Game Y** or Game Z.
Ⓒ Most students like **Game Y**.
Ⓓ Most students don't like **Game X**.

9. **Tom and Bob conducted two surveys about their co-workers' favorite hobbies. What inference can you make concerning their co-workers' favorite hobbies?**

Co-worker sample	Play Golf	Play Video Games	Fishing	Total
#1	125	80	126	331
#2	118	83	130	331

Ⓐ Most of their coworkers spend a lot of money.
Ⓑ Most of their coworkers spend a lot of time inside.
Ⓒ Most of their coworkers spend a lot of time outside.
Ⓓ Most of their coworkers live in big houses.

10. **Bill and Jill conducted two surveys about students' favorite card games. What inference can you make concerning the students' favorite card games?**

Student samples	Hearts	Go fish	Spades	Total
#1	14	90	11	115
#2	10	88	14	122

- Ⓐ Most students like speed.
- Ⓑ Most students like hearts.
- Ⓒ Most students like spades.
- Ⓓ Most students like Go fish over hearts or spades.

11. **Moe and Lonnie conducted two surveys about students' favorite soda flavor. What inference can you make concerning the students' favorite soda flavor?**

Student samples	Strawberry	Orange	Rootbeer	Total
#1	10	11	88	109
#2	14	14	90	118

- Ⓐ Most students like strawberry.
- Ⓑ Most students like orange.
- Ⓒ Most students like rootbeer over strawberry or orange.
- Ⓓ Most students like chocolate.

12. **Billy and Larry conducted two surveys about students' favorite subject. What inference can you make concerning the students' favorite subject?**

Student samples	Math	English Language Arts	Science	Total
#1	111	111	111	333
#2	114	114	114	342

- Ⓐ Most students like science.
- Ⓑ Most students like math.
- Ⓒ The same number of students like the three subjects equally.
- Ⓓ Most students like English Language Arts.

13. Tom and Jerry conducted two surveys about students' favorite ice cream flavor. What inference can you make concerning the students' favorite ice cream flavor?

Student samples	Chocolate	Strawberry	Cookies-n-Cream	Total
#1	59	3	12	74
#2	61	4	15	80

Ⓐ Most students' like cookies-n-cream.
Ⓑ Most students like chocolate over strawberry or cookies-n-cream.
Ⓒ Most students' like strawberry.
Ⓓ Most students' like strawberry and cookies-n-cream.

14. Paul and Maggie conducted two surveys about students' favorite holiday. What inference can you make concerning the students' favorite holiday?

Student samples	Christmas	Thanksgiving	Easter	Total
#1	100	128	244	472
#2	98	129	250	477

Ⓐ Most students like Halloween.
Ⓑ Most students like Christmas.
Ⓒ Most students like Thanksgiving.
Ⓓ The students like Easter over Christmas or Thanksgiving.

15. Harper wants to conduct two surveys at his school, where 1,600 students attend. For the first survey, he wants to find out what types of cell phones the students in his school use. Of those students, he wants to survey them to determine their favorite apps. Which of the following is the best method for selecting a random sample?

Ⓐ Select 20 students from each first period class.
Ⓑ Select random students as they enter or exit school.
Ⓒ Select students based on how many text messages they make each month.
Ⓓ Select all of the students.

16. You want to take a survey of 250 people in the mall in one day. You want to create a simple random sample of 100 people. Which method is most likely to result in a simple random sample?

Circle the correct answer choice.

Ⓐ Assign each person a number, then select every prime number.
Ⓑ Select the first 100 people to complete the survey.
Ⓒ Assign each person a number, then select 100 random numbers.
Ⓓ Select 100 people close to the bookstore.

17. Your school is thinking about starting a drama club but they want to know if students like the idea. To sample the population, the principal is going to survey every 10th student that walks into the school on Monday morning starting with the first student.

Circle the method of sampling this describes.

Ⓐ Simple Random Sampling
Ⓑ Systematic Sampling
Ⓒ Convenience Sampling

CHAPTER 5 → Lesson 5: Predicting Using Probability

1. **Which of the following represents the sample space for flipping two coins?**

   Ⓐ {HH, TT}
   Ⓑ {H, T}
   Ⓒ {HH, HT, TH, TT}
   Ⓓ {HH, HT, TT}

2. **Which of the following experiments would best test the statement, "The probability of a coin landing on heads is 1/2."?**

   Ⓐ Toss a coin 1,000 times, and record the results.
   Ⓑ Toss a coin twice to see if it lands on heads one out of those two times.
   Ⓒ Toss a coin until it lands on heads and record the number of tries it took.
   Ⓓ Toss a coin twice, if it doesn't land on heads exactly once, the theoretical probability is false.

3. **Which of the following results is most likely from rolling a six-sided die?**

   Ⓐ Rolling an odd number
   Ⓑ Rolling an even number
   Ⓒ Rolling a number from 1 to 3
   Ⓓ All of the above are equally likely.

4. **Sandy flipped a coin 40 times. Her results are 75% heads and 25% tails. What is the difference between the actual results and the expected results?**

   Ⓐ 20%
   Ⓑ 50%
   Ⓒ 25%
   Ⓓ 10%

5. **Maggie rolled a pair of four sided dice 10 times. The results are 30% side 1, 20% side 2, 20% side 3, 30% side 4. What is the difference between the results and the expected results for all four sides?**

   Ⓐ 25% side 1,
   25% side 2,
   25% side 3,
   25% side 4

   Ⓑ 5% side 1,
   5% side 2,
   5% side 3,
   5% side 4

   Ⓒ 5% side 1,
   25% side 2,
   25% side 3,
   5% side 4

   Ⓓ 4% side 1,
   4% side 2,
   4% side 3,
   4% side 4

6. **Maggie rolls two pairs of four sided dice 10 times. The results were 30% side 1, 20% side 2, 20% side 3, 30% side 4. What were the actual results and expected results?**

   Ⓐ Results: 12 side 1, 8 side 2, 8 side 3, 12 side 4 ...Expected Results: 10 side 1, 10 side 2, 10 side 3, 10 side 4
   Ⓑ Results: 6 side 1, 4 side 2, 4 side 3, 6 side 4 ...Expected Results: 5 side 1, 5 side 2, 5 side 3, 5 side 4
   Ⓒ Results: 6 side 1, 4 side 2, 4 side 3, 6 side 4 ...Expected Results: 10 side 1, 10 side 2, 10 side 3, 10 side 4
   Ⓓ Results: 10 side 1, 10 side 2, 10 side 3, 10 side 4 ...Expected Results: 12 side 1, 8 side 2, 8 side 3, 12 side 4

7. **Juliana was expected to make 80% of her first serves in the tennis match. She made half of 60 first serves in the match. What were the results and expected results?**

   Ⓐ Results: 30 first serves. Expected Results: 48 first serves
   Ⓑ Results: 20 first serves. Expected Results: 48 first serves
   Ⓒ Results: 25 first serves. Expected Results: 48 first serves
   Ⓓ Results: 35 first serves. Expected Results: 48 first serves

8. Which of the following represents the sample space for rolling a pair of four-sided dice?

Ⓐ (1,1) (1,2) (1,3) (1,4) (2,2) (2,3) (2,4) (3,3) (3,4) (4,4)
Ⓑ (1,2) (1,3) (1,4) (2,3) (2,4) (3,4)
Ⓒ (1,2) (1,3) (1,4) (2,3) (2,4)(3,4) (4,4)
Ⓓ (1,2) (1,3) (1,4) (2,3) (2,4)

9. Lea is playing a carnival game and has a chance to win a large stuffed teddy bear. In order to win, she has to guess the color of the cube she will pick out of a box. The box is shown below.

"Pick the Cube" Carnival Game

[Figure: a box containing 20 squares of various colors — black, white, and gray cubes scattered inside]

Based on the colors and number of blocks, what is Lea's chance of winning the game if she says that she will pick a black cube out of the box?

Ⓐ 60%
Ⓑ 10%
Ⓒ 90%
Ⓓ 40%

10. Ms. Green is passing out snacks during her tutoring session. She has 7 bags of chips, 13 candy bars, 16 fruit snacks, and 10 lollipops. If one of Ms. Green's students randomly selects a snack, what is the probability that he will select a fruit snack? Round your answer to the nearest tenth of a percent.

Ⓐ 34.8%
Ⓑ 65.2%
Ⓒ 16.0%
Ⓓ 84.0%

11. Mark the correct probabilities for each event.

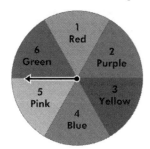

	$\frac{1}{6}$	$\frac{1}{3}$	$\frac{1}{2}$
The probability you spin an odd number.	○	○	○
The probability you spin a 3.	○	○	○
The probability you spin a blue.	○	○	○
The probability you spin a red or yellow.	○	○	○

12. What is the probability that the spinner will stop on the #3 sector?

Circle the correct answer choice.

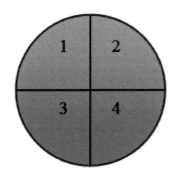

Ⓐ 1.3

Ⓑ $\frac{1}{2}$

Ⓒ $\frac{1}{3}$

Ⓓ $\frac{1}{4}$

CHAPTER 5 → Lesson 6: Understanding Probability

1. Mary has 3 red marbles and 7 yellow marbles in a bag. If she were to remove 2 red and 1 yellow marbles and set them aside, what is the probability of her pulling a yellow marble as her next marble?

   Ⓐ $\dfrac{1}{6}$

   Ⓑ $\dfrac{1}{7}$

   Ⓒ $\dfrac{7}{10}$

   Ⓓ $\dfrac{6}{7}$

2. John has a deck of cards (52 cards). If John removes a number 2 card from the deck, what is the probability that he will pick a number 2 card at random?

   Ⓐ 3 out of 51
   Ⓑ 4 out of 51
   Ⓒ 26 out of 51
   Ⓓ 30 out of 51

3. Maggie has a bag of coins (8 nickels, 6 quarters, 12 dimes, 20 pennies). If she picks a coin at random, what is the probability that she will pick a quarter?

   Ⓐ 2 out of 15
   Ⓑ 3 out of 23
   Ⓒ 3 out of 50
   Ⓓ 5 out of 46

4. Mark has a box of bills (12 ones, 8 tens, 21 twenties, 30 fifties). If he picks a bill at random, what is the probability that he will pick a ten?

   Ⓐ 8 out of 71
   Ⓑ 8 out of 100
   Ⓒ 10 out of 71
   Ⓓ 7 out of 71

5. Moe has a bowl of nuts (14 pecans, 8 walnuts, 28 almonds, 33 peanuts). If he picks a nut at random, what is the probability that he will pick a peanut?

    Ⓐ 33 out of 70
    Ⓑ 33 out of 80
    Ⓒ 33 out of 100
    Ⓓ 33 out of 83

6. Xavier has a bowl of nuts (14 pecans, 8 walnuts, 28 almonds, 33 peanuts). If he picks out all the pecans, what is the probability that he will pick a walnut at random?

    Ⓐ 8 out of 83
    Ⓑ 8 out of 69
    Ⓒ 8 out of 100
    Ⓓ 8 out of 70

7. Tim has a box of chocolates with the following flavors: 24 cherry, 26 caramel, 20 fudge, and 16 candy. If he picks out two of each type of chocolate, what is the probability that he will pick a cherry chocolate at random?

    Ⓐ 11 out of 35
    Ⓑ 4 out of 13
    Ⓒ 11 out of 50
    Ⓓ 11 out of 39

8. Clarissa has a box of chocolates with the following flavors: 24 cherry, 26 caramel, 20 fudge, and 16 taffy. If she picks out all the taffy, what is the probability that she will pick a fudge chocolate at random?

    Ⓐ 2 out of 7
    Ⓑ 10 out of 43
    Ⓒ 1 out of 5
    Ⓓ 8 out of 35

9. Karen has a box of chocolates with the following flavors: 24 cherry, 26 caramel, 20 fudge, and 16 taffy. If she removes half of the cherry and fudge chocolates from the box, what is the probability that she will pick a taffy chocolate at random?

    Ⓐ 1 out of 4
    Ⓑ 8 out of 43
    Ⓒ 4 out of 25
    Ⓓ 16 out of 43

10. **The table below shows the types of fruits Jessie's mom purchased from the grocery store.**

Type of Fruit	Number of Fruits
Apple	6
Orange	4
Pear	2
Peaches	3

If Jessie grabs one of the fruits without looking, what is the probability that he will NOT pick a peach?

Ⓐ $\dfrac{2}{5}$

Ⓑ $\dfrac{3}{15}$

Ⓒ $\dfrac{4}{5}$

Ⓓ $\dfrac{2}{3}$

11. **The table shows the types of books five friends like. One friend is chosen at random. Identify the outcome of the event. Event: The friend that likes non-fiction.**

Circle the outcome of the event shown below.

Favorite Types of Books

Ⓐ Friend - Type
Ⓑ Sarah - Fiction
Ⓒ John - Non-Fiction
Ⓓ Billy - Fantasy
Ⓔ Raj - Mystery
Ⓕ Kavi - Romance

12. When one number cube is rolled, the following six outcomes are possible.

Identify the outcomes for each event.

Fill in the blanks given in the table with the correct outcomes.

Event	Outcome
The number cube comes up odd.	1, 3, 5
The number cube comes up even.	
The number cube comes up greater than 3.	
The number cube comes up less than or equal to 5.	

CHAPTER 5 → Lesson 7: Using Probability Models

1. **Sara rolls two dice, one black and one yellow. What is the probability that she will roll a 3 on the black die and a 5 on the yellow die?**

   Ⓐ $\dfrac{1}{6}$

   Ⓑ $\dfrac{1}{12}$

   Ⓒ $\dfrac{2}{15}$

   Ⓓ $\dfrac{1}{36}$

2. **Which of the following represents the probability of an event most likely to occur?**

   Ⓐ 0.25
   Ⓑ 0.91
   Ⓒ 0.58
   Ⓓ 0.15

3. **Which of the following is not a valid probability?**

   Ⓐ 0.25

   Ⓑ $\dfrac{1}{5}$

   Ⓒ 1

   Ⓓ $\dfrac{5}{4}$

4. **Tom tosses a coin 12 times. The coin lands on heads only twice. If Tom tosses the coin one more time, what is the probability that the coin will land on heads?**

   Ⓐ 40%
   Ⓑ 50%
   Ⓒ 60%
   Ⓓ 15%

5. Joe has 5 nickels, 5 dimes, 5 quarters, and 5 pennies in his pocket. Six times, he randomly picked a coin from his pocket and put it back. Joe picked a dime every time. If he randomly picks a coin from his pocket again, what is the probability the coin will be a dime?

   (A) 33%
   (B) 100%
   (C) 24%
   (D) 25%

6. Jim rolls a pair of six-sided dice five times. He rolls a pair of two's five times in a row. If he rolls the dice one more time, what is the probability he will roll a pair of fours?

   (A) 1 out of 21
   (B) 1 out of 36
   (C) 1 out of 24
   (D) 1 out of 6

7. Bob rolls a six-sided die and flips a coin five times. He rolls a three and flips the coin to tails five times in a row. If he rolls the die and flips the coin one more time, what is the probability he will roll a three and flip the coin on tails?

   (A) 1 out of 6
   (B) 1 out of 10
   (C) 1 out of 12
   (D) 1 out of 7

8. Mia rolls a pair of six-sided dice and flips two coins ten times. She rolls a pair of threes and flips the coins to tails ten times in a row. If she rolls the dice and flips the coins one more time, what is the probability she will roll all fives and flip the coins on heads?

   (A) 1 out of 60
   (B) 1 out of 144
   (C) 1 out of 128
   (D) 1 out of 64

9. Sophia wants to select a pair of shorts from Too Sweet Clothing Store. The store has 2 different colors of shorts (black (B) and green (G)) available in sizes of small (S), medium (M), and large (L). If Sophia grabs a pair of shorts without looking, which sample space shows the different types of shorts she could select?

   (A) {BS, BM, BL, GS, GM, GL}
   (B) {BB, BG, SS, SM, SL}
   (C) {BG, SM, SL}
   (D) {BS, BL, GS, GL}

10. Eric's Pizza Shop is offering a sale on any large two-topping pizza for $8.00. Customers have a choice between thin crust (T) or pan pizza (Z), one meat topping [pepperoni(P), Italian sausage (I), or ham (H)], and one vegetable topping [green peppers (G), onions (O), or mushrooms (M)]. Which sample space shows the number of possible combinations customers can have if they select a thin crust pizza?

Ⓐ {TPG, TIO, THM}
Ⓑ {TPG, TPO, TPM, TIG, TIO, TIM, THG, THO, THM}
Ⓒ {TPG, TPO, TIG, TIO, THG, THO}
Ⓓ {TPG, TPO, TPM, TIG, TIO, TIM, THG, THO, THM, PPG, PPO, PPM, PIG, PIO, PIM, PHG, PHO, PHM}

11. There is a bag of tiles with 5 black tiles, 5 white tiles, and 5 blue tiles. You are going to choose one tile at random. Are the outcomes of this event equally likely?

Circle the correct answer.

Ⓐ Yes

Ⓑ No

12. You are playing a game using this spinner. You get one spin on each turn. Which list shows a complete probability model for the spinner?

Circle the correct answer.

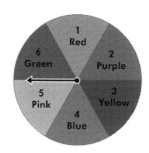

Ⓐ P(even number) = $\frac{3}{6}$, P(odd number) = $\frac{3}{6}$

Ⓑ P(red) = $\frac{1}{6}$, P(blue) = $\frac{1}{6}$

Ⓒ P(number <4) = $\frac{3}{6}$

Ⓓ P(yellow or green) = $\frac{2}{6}$, P(red or blue) = $\frac{2}{6}$

CHAPTER 5 → Lesson 8: Probability Models from Observed Frequencies

1. Felix flipped a coin 8 times and got the following results: H, H, T, H, H, T, T, H. If these results were typical for that coin, what are the odds of flipping a heads with that coin?

   Ⓐ 3 out of 5
   Ⓑ 5 out of 8
   Ⓒ 3 out of 8
   Ⓓ 1 out of 2

2. Bridgette rolled a six-sided die 100 times to test the frequency of each number's appearing. According to these statistics, how many times should a 2 be rolled out of 50 rolls?

Number	Frequency
1	18%
2	20%
3	16%
4	11%
5	18%
6	17%

   Ⓐ 10 times
   Ⓑ 20 times
   Ⓒ 12 times
   Ⓓ 15 times

3. Randomly choosing a number out of a hat 50 times resulted in choosing an odd number a total of four more times than the number of times an even number was chosen. How many times was an even number chosen from the hat?

   Ⓐ 27 times
   Ⓑ 21 times
   Ⓒ 29 times
   Ⓓ 23 times

4. 8 out of the last 12 customers at Paul's Pizza ordered pepperoni pizza. According to this data, what is the probability that the next customer will NOT order pepperoni pizza?

   Ⓐ 1 out of 3
   Ⓑ 4 out of 5
   Ⓒ 1 out of 2
   Ⓓ 2 out of 5

5. Susan is selling cookies for a fundraiser. Out of the last 20 people she asked, 10 people bought 1 box of cookies, 5 bought more than 1 box, and 5 bought none. Based on this data, what is the probability that the next person she asks will buy at least 1 box of cookies?

Ⓐ 1 out of 3
Ⓑ 2 out of 5
Ⓒ 4 out of 5
Ⓓ 3 out of 4

6. William has passed 9 out of his last 10 tests in Spanish class. Based on his past history, what is the probability that he will NOT pass the next test?

Ⓐ 10%
Ⓑ 25%
Ⓒ 15%
Ⓓ 8%

7. Travis has scored goals in 7 of his last 9 soccer games. At this rate, what is the probability that he will score in his next game? Round the nearest percent.

Ⓐ 70%
Ⓑ 17%
Ⓒ 78%
Ⓓ 53%

8. Gabe's free throw percentage for the season has been 80%. Based on this, if he has 5 free throws in the next game, how many is he likely to miss?

Ⓐ 0
Ⓑ 1
Ⓒ 2
Ⓓ 3

9. York and his partner have won the doubles tennis tournament three out of the last four years. According to this record, what is the probability they will win it again this year?

Ⓐ 3 out of 4
Ⓑ 1 out of 3
Ⓒ 1 out of 2
Ⓓ 4 out of 5

10. Robbie runs track. His finishes for his last 6 events were: 1st, 3rd, 2nd, 5th, 4th, 2nd. Based on these results, what is the probability he will finish in the top 3 of his next event?

Ⓐ 4 out of 5
Ⓑ 2 out of 3
Ⓒ 1 out of 2
Ⓓ 3 out of 4

11. The table shows observed frequencies of spinning a spinner with 4 numbers on it (1, 2, 3, 4).

	Experiment Table				
Outcome	1	2	3	4	50 total trials
Frequency	6	11	19	14	

What is the observed probability of spinning a 4?

Circle the correct answer choice.

Ⓐ $P(4) = \dfrac{2}{7}$

Ⓑ $P(4) = 14$

Ⓒ $P(4) = \dfrac{4}{14}$

Ⓓ $P(4) = \dfrac{7}{25}$

12. The table shows observed frequencies of rolling a six sided die. Mark the probabilities of different outcomes given in the table.

	Experiment Table						
Outcome	1	2	3	4	5	6	50 total trials
Frequency	7	9	10	8	9	7	

	$\dfrac{24}{50}$	$\dfrac{26}{50}$
Probability of rolling greater than or equal to 4	○	○
Probability of rolling less than 4	○	○
Probability of rolling an even number	○	○
Probability of rolling an odd number	○	○

CHAPTER 5 → Lesson 9: Find the Probability of a Compound Event

1. The following tree diagram represents Jane's possible outfits:

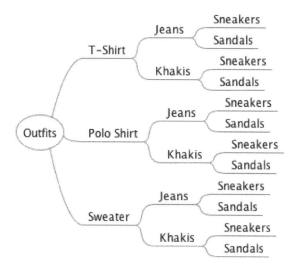

How many different outfits can Jane make based on this diagram?

Ⓐ 2
Ⓑ 12
Ⓒ 16
Ⓓ 4

2. The following tree diagram represents Jane's possible outfits:

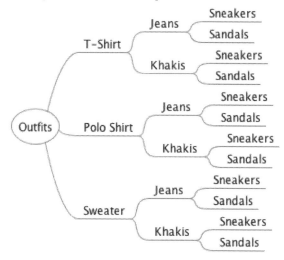

If Jane randomly selects an outfit, what is the probability she will be wearing Jeans AND Sneakers?

(A) $\dfrac{1}{12}$

(B) $\dfrac{1}{3}$

(C) $\dfrac{1}{4}$

(D) $\dfrac{1}{2}$

3. Paul, Jack, Tom, Fred, and Sam are competing in the long jump. If they win the top five spots, how many ways could they be arranged in the top five spots?

(A) 15 ways
(B) 60 ways
(C) 120 ways
(D) 3,125 ways

4. Mona is about to roll a pair of four-sided dice. What is the probability she will roll a one and a two or a one and a three?

(A) 1 out of 8
(B) 1 out of 16
(C) 1 out of 12
(D) 1 out of 4

5. Tim rolls a pair of six-sided dice. What is the probability he will roll doubles?

(A) 6 out of 21
(B) 1 out of 6
(C) 1 out of 4
(D) 2 out of 21

6. The triple jump competition is close. Joe, Damon, Sam, and Chris have a shot at first place. If two of the four tie for first place and the other two tie for second place, how many ways could they be arranged in the top two spots?

   Ⓐ 6 ways
   Ⓑ 2 ways
   Ⓒ 3 ways
   Ⓓ 8 ways

7. Sam is about to flip three coins. What is the probability he will flip all of the coins to heads?

   Ⓐ 1 out of 4
   Ⓑ 1 out of 6
   Ⓒ 1 out of 8
   Ⓓ 1 out of 2

8. Jona rolls one six-sided die and one four-sided die. What is the probability she will not roll a 2 or a 3 on either die?

   Ⓐ 2 out of 3
   Ⓑ 1 out of 3
   Ⓒ 1 out of 4
   Ⓓ 3 out of 4

9. Elsie rolled three four-sided dice. What is the probability she will roll one even and two odds?

   Ⓐ 1 out of 3
   Ⓑ 3 out of 8
   Ⓒ 1 out of 4
   Ⓓ 5 out of 8

10. What are the central tendencies of the following data set? (round to the nearest tenth)

    {21, 21, 22, 23, 25, 27, 28, 31, 34, 34, 34, 37}

    Ⓐ Mean: 28.1, Median: 27.5, Mode: 34
    Ⓑ Mean: 28.5, Median: 26, Mode: none
    Ⓒ Mean: 27.5, Median: 28, Mode: none
    Ⓓ Mean: 27, Median: 27.5, Mode: none

11. Are the events that make up the following compound events independent or dependent?

Fill in the blanks by writing the correct type against each compound event.

Player chooses a game piece.	dependent / independent
Spin the spinner. Then spin again.	
Pick a colored marble from a jar. Pick another marble from the jar.	
Roll a dice. Then roll a dice again.	

12. A jar is filled with 4 blue marbles, 2 yellow marbles, 5 red marbles, and 3 white marbles. Two marbles are chosen at random (one at a time) from the jar without replacement. What is the probability that a red marble and then a blue marble are chosen?

Write the answer in the box. Reduce the fraction to its simplest form and write it in the box given below.

End of Statistics & Probability

ANSWER KEY AND DETAILED EXPLANATION

Chapter 5
Statistics & Probability

Lesson 1: Mean, Median, and Mean Absolute Deviation

Question No.	Answer	Detailed Explanations
1	C	There is a strong positive correlation between the height and the weight because there is an upward trend in the weight as a person gets taller.
2	C	When analyzing the heights of the basketball players and soccer players, the average height of the basketball players is 5'4, and the average height of the soccer players is 5'2. Therefore, the average height of the basketball players is higher than the average height of the soccer players.
3	D	Remember: the mean represents the average of the values, which is calculated by adding all the values together then dividing by the number of values you added. In this case, $\frac{24 + 36 + 55 + 65}{2} = 45$, and $\frac{72 + 78}{2} = 75$. The difference between these two values is 30, as indicated.
4	A	Remember: A mean is the average of all the data presented. In this case, you have to average the temperatures, then replace the January temperature with 54 and recalculate the average, then take the difference. Another way of solving the problem is as follows: January temperature has increased by 54 - 24 = 30 degrees. So, this 30 degrees is to be distributed equally among all the six months. Therefore the average temperature increases by $\frac{30}{6} = 5$ degrees.
5	D	Remember: the mode of a set of data is the number that occurs most frequently. For this set of data, that number is 9.
6	C	Remember: Mean is the average, median is the middle-number when data is arranged numerically, and mode is the number that appears most often.

Question No.	Answer	Detailed Explanations
7	B	Mean absolute deviation is a measure of the distance between each data value and the mean, so it measures variability.
8	C	To find the mean, add the values, and divide that value by the amount of numbers in the data set. (1) The sum is $\frac{30}{4}$ (2) $\frac{30}{4} \div \frac{6}{1}$ = (3) $\frac{30}{4} \times \frac{1}{6} = \frac{30}{24}$ (the GCF of 30 and 24 is 6) (4) $\frac{5}{4}$
9	A	Mean is the average of a set of data; therefore, another word for mean is average.
10	A	To find the median, find the middle number in the data set. One way to find the median is to order the numbers from least to greatest and cross out the numbers until the middle is reached. Here, the middle number is $\frac{1}{3}$. Calculation: To compare fractions, we have to rewrite them with common denominator. Here LCM of 3 and 5 is 15. Rewrite all the fractions with 15 as the denominator. $\frac{4}{5} = \frac{4 \times 3}{5 \times 3} = \frac{12}{15}$. $\frac{1}{3} = \frac{1 \times 5}{3 \times 5} = \frac{5}{15}$. $\frac{1}{5} = \frac{1 \times 3}{5 \times 3} = \frac{3}{15}$. $\frac{2}{3} = \frac{2 \times 5}{3 \times 5} = \frac{10}{15}$. $\frac{3}{15} < \frac{5}{15} < \frac{10}{15} < \frac{12}{15}$. Therefore, when we arrange the data in ascending order, we get $\frac{1}{5}, \frac{1}{3}, \frac{1}{3}, \frac{2}{3}, \frac{4}{5}$. Median is 3rd score, which is $\frac{1}{3}$.
11	C	Mode is the value that appears most in a set of data. The list has a mode of 1 since that number appears most (3 times).
12	A	Remember: Mean is the average, median is the middle number when data is arranged numerically, and mode is the number that appears most often. Since mode is the value that appears most in a set of data, the original list has a mode of 14 because it is the number that appears most (2 times). Find the mean by adding the set of values and dividing by the amount of numbers in the set of data: (1) The sum of 6,8,1,5,11,14 and 14 = 59 (2) $\frac{59}{7}$ = 8.4 To find the median, find the middle number in the data set. One way to find the median is to order the numbers from least to greatest and cross out the numbers until the middle is reached. Here, the middle number is 8.

Question No.	Answer	Detailed Explanations
13	A	Find the mean by adding the set of values and dividing by the amount of numbers in the set of data: (1) The sum of 0.3, 1.2, 2.5, 4.3 = 8.3 (2) $\frac{8.3}{4}$ = 2.1 Therefore, the mean is 2.1.
14	C	To find the median, find the middle number in the data set. One way to find the median is to order the numbers from least to greatest and cross out the numbers until the middle is reached. Here, the middle numbers are 1.2 and 2.5. Add the numbers together and divide by 2. (1) 1.2 + 2.5 = 3.7 (2) 3.7 ÷ 2 = 1.85, which rounds to 1.9. Therefore, the median is 1.9
15	C	To find the mean absolute deviation, (1) find the mean, (2) find the difference between each data value and the mean (since we are interested in deviation only, ignore the negative sign), and (3) average the differences. Here, the mean is 3.4. The difference between the data values and mean are 1.4, 1.6, 3.6, 2.4 and 1.4. The average of the differences is 2.08.

16		

	Median	Mean
Data Set : 9, 8, 7, 4, 5	**7**	6.6
Data Set : 5, 8, 5, 9, 1, 2	5	**5**
Data Set : 1, 2, 8, 1, 3, 3	**2.5**	3

To find the mean, add all of the numbers together and divide by the number of items. To find the median, order the numbers from least to greatest and select the middle number. If there are two middle numbers, take the mean of the two numbers by adding them together and dividing by 2.

17	-3	The deviation of the minimum is the minimum - the mean. Here, minimum = 2, mean = 5 Therefore, the deviation of the minimum = 2 - 5 = -3

Lesson 2: Mean, Median, and Mode

Question No.	Answer	Detailed Explanations
1	B	Out of 16 scores, 13 of 16 are 4 or below. Since the majority of the scores are low, a person can infer that, overall, customers had a bad experience.
2	D	To answer this question, look at the results of the different dishes. Spaghetti: 7 Pizza: 13 Calzone: 9 Baked Ziti: 2 Since only 2 people said that baked ziti was their favorite food, it is the menu item the manager should remove from the menu.
3	A	The measures of central tendency are mean, which is the average of a set of data; median, which is the middle of a set of data, and mode, which is the value which appears most in a set of data.
4	A	To find the mean absolute deviation, (1) find the mean, (2) find the difference between each data value and the mean (since we are interested in deviation only, ignore the negative sign in the difference before taking average), and (3) average the differences. Here, the mean is 2.5. The differences between the data value and the mean are 1.5, 0.5, 0.5, and 1.5. The average is 1.
5	A	Since mode is the value that appears most in a set of data, the original list has two modes of 2 and 13 because each value appears the same number of times. Find the mean by adding the set of values and dividing by the amount of numbers in the set of data: (1) $2 + 2 + 3 + 4 + 6 + 8 + 9 + 10 + 13 + 13 + 16 + 17 = 103$ (2) $\frac{103}{12} = 8.6$ To find the median, find the middle number in the data set. One way to find the median is to order the numbers from least to greatest and cross out the numbers until the middle is reached. Here, the middle numbers are 8 and 9. Add the numbers together, and divide by 2. (1) $8 + 9 = 17$ (2) $17 \div 2 = 8.5$
6	D	Since mode is the value that appears most in a set of data, the original list has two modes of 11 and 23 since each value appears the same number of times. Find the mean by adding the set of values and dividing by the amount of numbers in the set of data: (1) $11 + 11 + 12 + 13 + 15 + 17 + 18 + 20 + 23 + 23 + 26 + 27 = 219$ (2) $\frac{216}{12} = 18$. To find the median, find the middle number in the data set. One way to find the median is to order the numbers from least to greatest and cross out the numbers until the middle is reached. Here, the middle numbers are 17 and 18. Add the numbers together and divide by 2. (1) $17 + 18 = 35$ (2) $35 \div 2 = 17.5$

Question No.	Answer	Detailed Explanations
7	D	Since mode is the value that appears most in a set of data, the original list has two modes of 31 and 54 since each value appears the same number of times. Find the mean by adding the set of values and dividing by the amount of numbers in the set of data: (1) The sum of 31, 31, 32, 33, 35, 37, 38, 41, 54, 54, 57, 58 = 501 (2) $\frac{501}{12}$ = 41.8 To find the median, find the middle number in the data set. One way to find the median is to order the numbers from least to greatest and cross out the numbers until the middle is reached. Here, the middle numbers are 37 and 38. Add the numbers together, and divide by 2. (1) 37 + 38 = 75 (2) 75 ÷ 2 = 37.5
8	A	Since mode is the value that appears most in a set of data, the original list has a mode of 64 since the number appears most (3 times). Find the mean by adding the set of values and dividing by the amount of numbers in the set of data: (1) The sum of 41, 41, 42, 43, 45, 47, 48, 51, 64, 64, 67, 64 = 617 (2) $\frac{617}{12}$ = 51.4 To find the median, find the middle number in the data set. One way to find the median is to order the numbers from least to greatest and cross out the numbers until the middle is reached. Here, the middle numbers are 47 and 48. Add the numbers together, and divide by 2. (1) 47 + 48 = 95 (2) 95 ÷ 2 = 47.5
9	B	Since mode is the value that appears most in a set of data, the original list has a mode of 54 since the number appears most (3 times). Find the mean by adding the set of values and dividing by the amount of numbers in the set of data: (1) The sum of 51, 51, 52, 53, 55, 57, 58, 61, 54, 54, 57, 54 = 657 (2) $\frac{657}{12}$ = 54.8 To find the median, find the middle number in the data set. One way to find the median is to order the numbers from least to greatest and cross out the numbers until the middle is reached. Here, the middle numbers are 54 and 54. Add the numbers together, and divide by 2. (1) 54 + 54 = 108 (2) 108 ÷ 2 = 54
10	A	Since mode is the value that appears most in a set of data, the original list has a mode of 61 since the number appears most (3 times). Find the mean by adding the set of values and dividing by the amount of numbers in the set of data: (1) The sum of 61, 61, 52, 53, 65, 67, 58, 61, 64, 64, 57, 54 = 717 (2) $\frac{717}{12}$ = 59.8 To find the median, find the middle number in the data set. One way to find the median is to order the numbers from least to greatest and cross out the numbers until the middle is reached. Here, the middle numbers are 61 and 61. Add the numbers together, and divide by 2. (1) 61 + 61 = 122 (2) 122 ÷ 2 = 61

Question No.	Answer	Detailed Explanations
11	C	Since mode is the value that appears most in a set of data, the original list has two modes -- 71 and 74 -- since those numbers appear most (3 times). Find the mean by adding the set of values and dividing by the amount of numbers in the set of data: (1) The sum of 71, 71, 62, 63, 75, 77, 68, 71, 74, 74, 67, 74 = 847 (2) $\frac{847}{12}$ = 70.6 (when rounded to the nearest tenth). To find the median, find the middle number in the data set. One way to find the median is to order the numbers from least to greatest and cross out the numbers until the middle is reached. Here, the middle numbers are 71 and 71. Add the numbers together and divide by 2. (1) 71 + 71 = 142 (2) 142 ÷ 2 = 71
12	A	Since mode is the value that appears most in a set of data, the original list has two modes -- 71 and 74 -- since those numbers appear most (2 times). Find the mean by adding the set of values and dividing by the amount of numbers in the set of data: (1) The sum of 61, 71, 52, 63, 65, 77, 58, 71, 64, 74, 57, 74 = 787 (2) $\frac{787}{12}$ = 65.6 To find the median, find the middle number in the data set. One way to find the median is to order the numbers from least to greatest and cross out the numbers until the middle is reached. Here, the middle numbers are 64 and 65. Add the numbers together, and divide by 2. (1) 64 + 65 = 129 (2) 129 ÷ 2 = 64.5
13	B	Since mode is the value that appears most in a set of data, the original list has a mode of 44 since that number appears most (3 times). Find the mean by adding the set of values and dividing by the amount of numbers in the set of data: (1) The sum of 21, 21, 22, 23, 35, 37, 38, 41, 44, 44, 47, 44 = 417 (2) $\frac{417}{12}$ = 34.8 To find the median, find the middle number in the data set. One way to find the median is to order the numbers from least to greatest and cross out the numbers until the middle is reached. Here, the middle numbers are 37 and 38. Add the numbers together, and divide by 2. (1) 37 + 38 = 75 (2) 75 ÷ 2 = 37.5

Question No.	Answer	Detailed Explanations
14	C	Selecting 108 is the result of finding the sum of the six values in the table and dividing by 6. Selecting 110 results from choosing the mean per day the company needs to make a profit. A mean of 222 results from a calculation error when adding or subtracting the sum of the values and the needed number of visitors. The mean is the average of the values in a set of data. In order for Kelli's Ice Cream Shop to have a mean of 110 per day, multiply 110 times the number of days. 110 x 7 = 770. Kelli's Ice Cream Shop needs 770 visitors per week to make a profit. Add the number of visitors from the table: 63 + 77+ 121 + 96 + 137 + 154 = 648. Subtract the sum from 770. 770 - 648 = 122 Therefore, Kelli's needs 122 visitors on Saturday in order to have a mean of 110 visitors per day for the week and to turn a profit.
15	C	Since mode is the value that appears most in a set of data, the original list has no mode since each value appears the same number of times. If 90 is added to the list, the score would appear twice. Therefore, the mode would be 90. We cannot say anything about mode i.e. whether mode has increased or decreased. So, option (A) is incorrect. The mean would neither increase nor decrease. The mean of the original list is 90, so adding a value of 90 would not change the mean. In the original list, the middle value of the numbers is 89. By adding 90 to the list, there will be two middle values, 89 and 90. The sum of the two middle values is 179. Divide the sum by 2. The quotient, 89.5, is the new median. Therefore, the median will increase by adding 90 to the list.

16

	Mean	Median	Mode
6	●		
5.5		●	
5			●

To find the mean, add all of the numbers together and divide by the number of items. To find the median, order the numbers from least to greatest and select the middle number. If there are two middle numbers, take the mean of the two numbers by adding them together and dividing by 2. Mode is the number that occurs most often.

Mean $= \dfrac{4 + 5 + 5 + 6 + 7 + 9}{6} = \dfrac{36}{6} = 6$

Median $= \dfrac{(3rd\ score + 4th\ score)}{2} = \dfrac{(5+6)}{2} = \dfrac{11}{2} = 5.5$

Mode = 5 (the score 5 has appeared twice)

Question No.	Answer	Detailed Explanations
17	40	Interquartile range (IQR) is the difference between the third and the first quartiles in descriptive statistics. So, you order the numbers from least to greatest: 16.6 22.4 32.6 49.3 51.8 68.2 72.6 98.5 99.8. There are 9 scores. 51.8 is the 5th score, which is the median. To find the first quartile and third quartile, include 51.8 in both the first half and the second half.

First half : 16.6, 22.4, 32.6, 49.3, 51.8. First quartile is the middle score of the first half, which is, 32.6.

Second half : 51.8, 68.2, 72.6, 98.5, 99.8. Third quartile is the middle score of the second half , which is, 72.6.

Therefore, Interquartile range (QR) = 72.6 - 32.6 = 40.

Lesson 3: Sampling a Population

Question No.	Answer	Detailed Explanations
1	C	A group of randomly selected students in the hallways would produce the least amount of bias because it is unlikely for assumptions to be made or factors that influence the data to be present. For example, students in the 8th grade may be taller than other students in other grades, skewing the data toward a higher average. A similar assumption can be made about students on the basketball team. Joe and Mary would already have an idea of the height of their friends.
2	A	A random, representative group represents the people to survey in a population because each person in the population has an equal chance of being included and there is less of a chance of bias altering the results of the survey.
3	C	As the size of the sample (people surveyed) increases, the results become more accurate. Therefore, increasing the sample size increases the reliability of the results.
4	D	Surveying everyone would produce reliable results because there would be specific data from each person; however, due to the length of time needed, among other reasons, surveying everyone may not be possible (depending on the reasons for the survey).
5	B	A sample size gets information from a small group from the population. Since the sample size is supposed to represent the population at large, it's important to know the parameters of the sample size in order to determine how reliable the results are for a given survey.
6	B	An assumption can be made that students on the basketball team are likely to be taller than other students at the school, which can influence the results (make it appear that the average height of the general school population is higher than it actually is). Therefore, calculating the average height of the basketball team would most likely produce the most amount of bias.
7	B	Multiple choice questions are useful for statistical results because they can represents different amounts of data that can be separated into categories and sub-categories based on the type of survey and the intended outcome.
8	C	Bias in collecting data can be caused by using questions where there are likely to be assumptions made or factors that influence the data. A survey that asks questions that cause individuals to answer in a certain way makes assumptions and influences the outcome of the data. Therefore, using that type of survey does not avoid bias.

Question No.	Answer	Detailed Explanations
9	B	When analyzing the data, out of the 10 responses, none of the responses have been less than 6. Therefore, a surprising response from the next person would be a 5 since it is outside the range.
10	B	When analyzing the data, out of the 10 responses, all are within the range of 6 to 8. Therefore, a surprising response from the next person would be a 10 since it is outside of the range.
11	A	Bias in collecting data can be caused by using questions where there are likely to be assumptions made or factors that influence the data. An anonymous survey would encourage individuals to answer questions openly without worry about the implications of their answers. Therefore, an anonymous survey is the best way to avoid receiving biased data.
12	A	Multiple choice questions are useful for statistical results because they can represent different amounts of data that can be separated into categories and sub-categories based on the type of survey and the intended outcome. The optional comments section gives an individual the opportunity to make open observations (without bias). A survey with too many free response questions would be hard to analyze and personal questions invite bias.
13	A	For collecting personal data, an anonymous survey would encourage individuals to answer questions openly without worry about the implications of their answers. It is also a way to avoid receiving bias data. Therefore, a receipt should have a link to an online survey where individuals can enter information about personal data.
14	A	In order to get a reliable, random sample based on the population (players who use the game), the surveyors should set up their survey at the end of a level or section of the game. Players are more likely to answer the survey before going to the next level. People are ready to play at the beginning of a game and ready to turn the game off near the end, so there is a good chance they won't complete the survey. A separate survey is likely to be ignored.
15	C	When choosing a sample to survey for data, the sample should be representative of the target population. Here, the target population is people most likely to purchase music online. These people are most likely to communicate with others about music online. Therefore, an online survey on music social networking websites would give them the best sample.

	Convenience Sample	Systematic Sample	Simple Random Sample
A person chooses every 5th person on a list of names starting with #1.	○	●	○
A person accepts the first 15 people to respond to a magazine ad.	●	○	○
A person picks names out of a hat.	○	○	●

Question No.	Answer	Detailed Explanations
16		(see table above)

Systematic sampling is a type of probability sampling method in which sample members from a larger population are selected according to a random starting point and a fixed, periodic interval.

Convenience sampling is a statistical method of drawing representative data by selecting people because of the ease of their volunteering or selecting units because of their availability or easy access.

A simple random sample is a subset of a statistical population in which each member of the subset has an equal probability of being chosen. A simple random sample is meant to be an unbiased representation of a group.

| 17 | 30 | If the representative sample is 10 then you would multiply the number of gray marbles by 5 to get the approximate number of gray marbles. 6 x 5 = 30. |

Lesson 4: Describing Multiple Samples

Question No.	Answer	Detailed Explanations
1	B	A group of randomly selected strangers in different places of a mall would produce the least amount of bias because there is unlikely to be assumptions made or factors that influence the data.
2	A	In the survey, 83 students selected science as their favorite subject while the other subjects had a combined number of 94 students. Therefore, an inference can be made that students prefer science over the other subjects.
3	A	In the sample, a combined 93 students chose action movies, 41 students chose drama, and 73 chose comedy. Therefore, an inference can be made that students prefer action movies to other types of movies.
4	A	In the sample, a combined 84 students chose strawberries, 38 students chose bananas and 63 chose blueberries. Therefore, an inference can be made that students prefer strawberries to other types of fruit.
5	A	In the sample, a combined 295 students chose basketball, 106 students chose soccer, and 46 students chose play tennis. Since more students chose basketball, an inference can be made that most students like basketball over soccer or tennis.
6	A	In the sample, a combined 198 students chose summer, 257 students chose fall, and 494 students chose spring. Since more students chose spring, an inference can be that most students prefer the Spring season over Summer or Fall.
7	D	In the sample, a combined 295 students chose pineapple, 238 students chose extra cheese, and 467 students chose pepperoni. Since more students chose pepperoni, an inference can be that most students prefer pepperoni over pineapple or extra cheese.
8	B	In the sample, a combined 95 students chose Game X, 11 students chose Game Y, and 8 students chose Game Z. Since more students chose Game X, an inference can be that most students like Game X over Game Y or Game Z.
9	C	In the sample, a combined 243 co-workers play golf, 163 co-workers chose video games, and 256 co-workers chose fishing. Since golf and fishing are outdoor activities, an inference can be made that most of Tom and Bob's co-workers spend a lot of time outside.

Question No.	Answer	Detailed Explanations
10	D	In the sample, a combined 178 students chose Go fish, 24 students chose hearts, and 25 students chose spades. Since more students chose Go fish, an inference can be that most students like Go fish over hearts or spades.
11	C	In the sample, a combined 24 students chose strawberry, 178 students chose rootbeer, and 25 students chose orange. Since more students chose rootbeer, an inference can be made that most students prefer rootbeer over strawberry or orange.
12	C	In the sample, a combined 225 students chose math, 225 students chose English and 225 students chose science. Since the students' answers reflect the same numbers for each subject, an inference can be made that the same number of students like the three subjects equally.
13	B	In the sample, a combined 120 students chose chocolate, 7 students chose strawberry and 27 students chose cookies-n-cream. Since more students chose chocolate, an inference can be made that most students like chocolate over strawberry or cookies-n-cream.
14	D	In the sample, a combined 198 students chose Christmas, 257 students chose Thanksgiving and 494 students chose Easter. Since more students chose Easter, an inference can be made that most students like Easter over Christmas or Thanksgiving.
15	B	A random, representative group represents the people to survey in a population because each person in the population has an equal chance of being included and there is less of a chance of bias altering the results of the survey. Therefore, Harper should select students as they enter and exit the school.
16	C	C is a simple random sample because A is systematic, B is convenience, and D is biased.
17	B	Systematic sampling is a type of probability sampling method in which sample members from a larger population are selected according to a random starting point and a fixed periodic interval. Hence, the sampling method shown above is Systematic sampling.

Lesson 5: Predicting Using Probability

Question No.	Answer	Detailed Explanations
1	C	Create a sample space by listing all the possible outcomes. For each coin flipped, it will land on heads or tails. Therefore, for two coins, there could be outcomes of (heads, heads), (heads, tails), (tails, heads), and (tails, tails).
2	A	When testing probability, larger samples (experiments) yield more accurate results. An experiment of 1,000 coin tosses could adequately test if a coin would land on its head $\frac{1}{2}$ of the time (about 500 of 1,000).
3	D	Since there are three even numbers (2, 4, 6) and three odd numbers (1, 3, 5) on a six-sided die, the probability is 3 out of 6 for rolling either an even or an odd number. Therefore, it is equally likely to roll an even or odd number. There is also a 3 out of 6 chance of rolling a number from 1 to 3.
4	C	The theoretical probability (expected outcome) of flipping a coin is 1 out of 2 because it will either land on heads or tails (those are the two possible outcomes). 1 out of 2 is equivalent to 50%. Subtract the expected outcome from the actual outcome 75% (75 - 50 = 25) and add the percent sign: 25%
5	B	The theoretical probability (expected outcome) of rolling each die is 1 out of 4 because it will either land on 1, 2, 3 or 4 (those are the four possible outcomes). One out of 4 is equal to 25%. Therefore, the difference between the expected results and the actual results is 5% for each side.
6	A	The theoretical probability (expected outcome) of rolling each die is 1 out of 4 because it will either land on 1, 2, 3 or 4 (those are the four possible outcomes). So, the expected outcome would be 10 for each side. The actual outcome was 40(0.30) for side one and side four and 40(.20) for side two and side three, which equals 12 side one, 8 side two, 8 side three, and 12 side four.
7	A	If Juliana made 60 serves, multiply 60 by the decimal form of 80% to find her expected first serve number. 60 × .80 = 48 first serves expected. If she only made half of the serves, divide 60 by 2. 60 ÷ 2 = 30 actual serves made.
8	A	Create a sample space by listing all the possible outcomes. Combining duplicate outcomes leaves 10 possible results: (1,1) (1,2) (1,3) (1,4) (2,2) (2,3) (2,4) (3,3) (3,4) (4,4)

Question No.	Answer	Detailed Explanations
9	D	There are 10 black cubes in the box and 25 total cubes. Therefore, Lea has a 10 out of 25 chance to pick a black cube. Multiply both 10 and 25 by 4 in order to find out her chances in percent form. 10 x 4 = 40 and 25 x 4 = 100. Since there is a 40 out of 100 chance to pick a black cube, Lea's chances of winning the game is 40%.
10	A	Since there are 16 fruit snacks out of a total of 46 total snacks, the probability of randomly picking a fruit snack is 16 out of 46. Divide 16 by 46, and then multiply by 100 in order to find the probability in percent form. 16 ÷ 46 = 0.3478 x 100 = 34.78%. Rounding to the nearest tenth means a probability of 34.8% that a student will randomly select a fruit snack.

11

	$\frac{1}{6}$	$\frac{1}{3}$	$\frac{1}{2}$
The probability you spin an odd number.			●
The probability you spin a 3.	●		
The probability you spin a blue.	●		
The probability you spin a red or yellow.		●	

(1) Total number of possible outcomes = 6. Number of favorable outcomes (you spin a odd number) = 3 (it may be 1, or 3 or 5).
Therefore Probability = (number of favorable outcomes) / (total number of possible outcomes) = 3/6 = 1/2
(2) Total number of possible outcomes = 6. Number of favorable outcomes (you spin the number 3) = 1
Therefore Probability = (number of favorable outcomes) / (total number of possible outcomes) = 1/6
(3) Total number of possible outcomes = 6. Number of favorable outcomes (you spin a blue) = 1
Therefore Probability = (number of favorable outcomes) / (total number of possible outcomes) = 1/6
(4) Total number of possible outcomes = 6. Number of favorable outcomes (you spin a red or yellow) = 2
Therefore Probability = (number of favorable outcomes) / (total number of possible outcomes) = 2/6 = 1/3

| 12 | D | The probability is the number of favorable outcomes over the number of outcomes in a sample space. Here, number of favorable outcomes = 1, number of outcomes in the sample space = 4. Therefore probability = $\frac{1}{4}$ |

Lesson 6: Understanding Probability

Question No.	Answer	Detailed Explanations
1	D	Mary originally had 10 marbles in her bag. When she removed 3 marbles, she had 7 remaining marbles. Six of those marbles are yellow. Probability is: (chance of successful outcome)/(total number of outcomes). Plug in the numbers, and solve. 6 chances to pick yellow out of 7 outcomes = $\frac{6}{7}$
2	A	There are four number 2 cards in a deck of 52. If one number 2 card is removed, that leaves three number 2 cards out of 51 cards. Probability is: (chance of successful outcome)/(total number of outcomes). Therefore, there is a 3 out of 51 chance that John will pick number 2 at random.
3	B	Maggie has a total of 46 coins, 6 of which are quarters. Probability is: $\frac{\text{chance of successful outcome}}{\text{total number of outcomes}}$. Therefore, there is a 6 out of 46 chance that Maggie will pick a quarter. This simplifies to 3 out of 23.
4	A	Mark has a total of 71 bills, 8 of which are tens. Probability is: $\frac{\text{chance of successful outcome}}{\text{total number of outcomes}}$. Therefore, there is an 8 out of 71 chance that Mark will pick a ten.
5	D	Moe has a total of 83 nuts, 33 of which are peanuts. Probability is: $\frac{\text{chance of successful outcome}}{\text{total number of outcomes}}$. Therefore, there is a 33 out of 83 chance that Moe will pick a peanut.
6	B	Xavier has a total of 83 nuts, 8 of which are walnuts. If he removes the pecans, he will have 69 nuts remaining (83 - 14 = 69). Probability is: $\frac{\text{chance of successful outcome}}{\text{total number of outcomes}}$. Therefore, there is an 8 out of 69 chance that Xavier will pick a walnut.
7	D	Tim had a total of 86 chocolates. If he removes 2 of each type of chocolate, he will have 22 cherry, 24 caramel, 18 fudge, and 14 candy remaining for a total of 78 chocolates. Probability is: $\frac{\text{chance of successful outcome}}{\text{total number of outcomes}}$ Therefore, there is a 22 out of 78 chance that Tim will pick a cherry chocolate. This reduces to 11 out of 39.

Question No.	Answer	Detailed Explanations
8	A	Clarissa had a total of 86 chocolates. If she removes all of the taffy, she will have 70 chocolates remaining. Probability is: $\frac{\text{chance of successful outcome}}{\text{total number of outcomes}}$ Therefore, there is a 20 out of 70 chance that Clarissa will pick a fudge chocolate. This reduces to 2 out of 7.
9	A	Karen had a total of 86 chocolates. If she removes $\frac{1}{2}$ of the cherry and fudge chocolates, she will have 12 cherry, 26 caramel, 10 fudge, and 16 taffy remaining for a total of 64 chocolates. Probability is: $\frac{\text{chance of successful outcome}}{\text{total number of outcomes}}$. Therefore, there is a 16 out of 64 chance that Karen will pick a taffy chocolate. This reduces to 1 out of 4.
10	C	Probability of Jessie picking up a peach is $\frac{3}{15}$, subtract that amount from $\frac{15}{15}$ (Probability of picking one fruit of any type $= 1 = \frac{15}{15}$). $\frac{15}{15} - \frac{3}{15} = \frac{12}{15}$, which in simplest form is $\frac{4}{5}$. Therefore, the probability that he will not pick a peach is $\frac{4}{5}$.
11	C	An event is a group of one or more possible outcomes. The outcome John satisfies the event of choosing Non-Fiction.
12		

Event	Outcome
The number cube comes up odd.	1, 3, 5
The number cube comes up even.	2,4,6
The number cube comes up greater than 3.	4,5,6
The number cube comes up less than or equal to 5.	1,2,3,4,5

The answers are :
1) The number cube comes up even - 2,4,6
2) The number cube comes up greater than 3 - 4,5,6
3) The number cube comes up less than or equal to 5 - 1,2,3,4,5
This answer requires the identification of odd, even, greater than 3 & less than or equal to 5 outcomes.

Lesson 7: Using Probability Models

Question No.	Answer	Detailed Explanations
1	D	Remember: In a compound and event, you can multiply the probabilities of each event in order to arrive at a final solution. There is a $\frac{1}{6}$ chance she will roll a 3 and a $\frac{1}{6}$ chance Sara rolls a 5. This means there is a $\frac{1}{6}(\frac{1}{6}) = \frac{1}{36}$ chance she will roll both. Another way to approach this problem is to create the sample space containing all possible combinations. (B, Y): (1,1), (1,2), (1,3), (1,4), (1,5), (1,6) (2,1), (2,2), (2,3), (2,4), (2,5), (2,6) (3,1), (3,2), (3,3), (3,4), (3,5), (3,6) (4,1), (4,2), (4,3), (4,4), (4,5), (4,6) (5,1), (5,2), (5,3), (5,4), (5,5), (5,6) (6,1), (6,2), (6,3), (6,4), (6,5), (6,6) From this list it is clear that only one out of the 36 total possible matches the (3,5) described.
2	B	Probability is defined as chance that an event will occur (a number between 0 and 1). The closer a probability is to 1, the more likely it is to occur. Drawing a number line with marks 0.1 distance apart can help determine the likelihood of an event occurring. Since 0.91 is close to 1, it represents an event most likely to occur.
3	D	Probability is defined as the chance that an event will occur (a number from 0 to 1). Since $\frac{5}{4}$ is a rational number greater than 1, it cannot represent a probability.
4	B	Theoretical probability is the (number of possible favorable outcomes)/(total number of outcomes). Each time Tom flips a coin, there is 1 possible favorable outcome (heads) out of 2 total possible outcomes. Therefore, the probability is $\frac{1}{2}$, which is equivalent to 50%.
5	D	Since there are the same amount of coins in each amount, the theoretical probability (expected outcome) of picking a coin is 1 out of 4 because Joe will either pick a penny, a nickel, a dime or a quarter. 1 out of 4 is equivalent to 25%.
6	B	For every roll of two 6-sided dice, there are 36 possible outcomes. Therefore, the probability of rolling a pair of fours is 1 out of 36.

Question No.	Answer	Detailed Explanations
7	C	For every roll of a 6-sided die and coin flip, there are 12 possible outcomes: {(1,h), (1,t), (2,h), (2,t), (3,h), (3,t), (4,h), (4,t), (5,h), (5,t), (6,h), (6,t)}. Therefore, the possibility of rolling a three and flipping tails is 1 out of 12.
8	B	For every roll of two 6-sided dice and two coin flips, there are 144 possible outcomes (36 x 4). Therefore, the probability of rolling two fives and flipping two heads is 1 out of 144.
9	A	The sample space of {BS, BM, BL, GS, GM, GL} is the correct answer. Since Sophia has a choice of 2 different pairs of shorts in 3 different combinations, the sample space should reflect choices of black or green shorts in sizes of small, medium, or large.
10	B	For each type of crust, there are 9 possible combinations. Therefore, the sample space {TPG, TPO, TPM, TIG, TIO, TIM, THG, THO, THM} represents the number of combinations if a customer chooses a thin crust pizza.
11	Yes	The outcomes are equally likely because the probability of choosing each color is $\frac{5}{15}$.
12	A	A probability model must add up to equal 1 to account for all possibilities.

Lesson 8: Probability Models from Observed Frequencies

Question No.	Answer	Detailed Explanations
1	B	Because 5 of the 8 times it was flipped, the result was heads, the probability of heads is 5 out of 8.
2	A	If 20% of the rolls result in a 2, then 20% of 50 rolls would be (0.20)(50) = 10 times.
3	D	An even distribution of the odds and evens would be 25 each. 4 more odds would mean 27 odds and 23 evens.
4	A	4 out of 12 customers did NOT order pepperoni. This simplifies to 1 out of 3.
5	D	15 out of the 20 people ordered 1 box or more. This reduces down to 3 out of every 4 people.
6	A	Only 1 time out of 10 did he NOT pass the test, so this is a 10% probability.
7	C	7 out of 9 is a 77.777...% probability, which rounds to 78%.
8	B	An 80% success rate means that he makes 4 out of 5 and misses only 1 out of every 5 free throws.
9	A	The fact that 3 of the last 4 years they won the tournament suggests that this year's probability of winning is that same 3 out of 4.
10	B	4 out of the last 6 results were in the top three, so this simplifies to a 2 out of 3 probability.
11	D	The probability of spinning a 4 is $\frac{14}{50}$, which simplified is $\frac{7}{25}$

	Question No.	Answer	Detailed Explanations

Question No.	Answer	Detailed Explanations
12		

	$\dfrac{24}{50}$	$\dfrac{26}{50}$
Probability of rolling greater than or equal to 4	●	
Probability of rolling less than 4		●
Probability of rolling an even number	●	
Probability of rolling an odd number		●

Add the frequencies together for each part of the outcome and divide that by the total number of trials.

(1) The probability of rolling greater than or equal to 4 means rolling a 4, 5, or 6. Therefore, the frequencies are 8 + 9 + 7 = 24 out of 50.
(2) The probability of rolling less than 4 means rolling a 1, 2, or 3. Therefore, the frequencies are 7 + 9 + 10 = 26 our of 50.
(3) The probability of rolling an even number means rolling a 2, 4 or 6. Therefore, the frequencies are 9 + 8 + 7 = 24 out of 50.
(4) The probability of rolling an odd number means rolling a 1, 3 or 5. Therefore, the frequencies are 7 + 10 + 9 = 26 out of 50.

Lesson 9: Find the Probability of a Compound Event

Question No.	Answer	Detailed Explanations
1	B	For the t-shirt, polo shirt and sweater, determine how many outfits Jane can make counting the number of jeans and khakis and sneakers and sandals she can match with her shirts. Therefore, there are 12 combinations available for the outfits Jane can make.
2	C	For the t-shirt, polo shirt and sweater, determine how many outfits Jane can make, counting the number of jeans and khakis and sneakers and sandals she can match with her shirts. Therefore, there are 12 combinations available for the outfits Jane can make. For 12 outfits, there are 3 outfits with jeans and sneakers. Therefore, there is a 3 out of 12 probability that Jane will wear jeans and sneakers. 3 out of 12 is $\frac{3}{12}$, which is $\frac{1}{4}$ in simplest form.
3	C	For each name in first place, there are 24 ways the other names can be arranged. $24 \times 5 = 120$ ways. Alternate Method : First spot can be arranged in 5 ways. After selecting the first spot, second spot can be arranged in 4 ways and so on for the third, fourth and fifth spot. So Top five spots can be arranged in $5 \times 4 \times 3 \times 2 \times 1 = 120$ ways.
4	D	For every roll of two 4-sided dice, there are 16 possible outcomes. Four of these outcomes would be favorable (1, 2), (2, 1), (1, 3), or (3, 1). So, the probability is 4 out of 16, or 1 out of 4.
5	B	For every roll of two 6-sided dice, there are 36 possible outcomes. Since there are 6 doubles combinations, the probability of rolling a double is 6 out of 36, or 1 out of 6.
6	A	There are 6 ways to arrange the players for the top two spots (combining duplicates): JDSC, JCDS, SCDJ, SJCD, CDSJ, SDJC.
7	C	For every three coin flips, there are 8 possible outcomes. 1 of those outcomes would contain all heads. So, the probability is 1 out of 8.
8	B	For every roll of a 6-sided die and a 4-sided die, there are 24 possible outcomes. Of these, 16 outcomes contain a 2 or a 3 (or both). This leaves 8 favorable outcomes. So the probability is 8 out of 24, or 1 out of 3.

Question No.	Answer	Detailed Explanations
9	B	For every roll of three 4-sided dice, there are 64 possible outcomes. 24 of these outcomes would contain one even and two odds. Therefore, the probability is 24 out of 64, or 3 out of 8.
10	A	To find the mean, add the numbers together, and divide by the amount of numbers in the data set. (1) 21 + 21 + 22 + 23 + 25 + 27 + 28 + 31 +34 + 34 + 34 + 37 = 337 (2) 337 ÷ 12 = 28.1 To find the median, find the middle number in the data set. One way to find the median is to order the numbers from least to greatest and cross out the numbers until the middle is reached. Here, the middle numbers are 27 and 28. Add the numbers together and divide by 2. (1) 28 + 27 = 55 (2) 55 ÷ 2 = 27.5 To find the mode, find the number that appears most in the data set. Looking at the numbers shows that the number 34 appears 3 times; therefore 34 is the mode.
11		In probability theory, independent events are events where the occurrence or non-occurrence of one event does not affect the probability of the occurrence of another event. Dependent events, on the other hand, are events where the occurrence or non-occurrence of one event affects the probability of the occurrence of another event.
		Based on the above definitions, the answers are :he answers are:
		Spin the spinner. Then spin again. - Independent
		Pick a colored marble from a jar. Pick another marble from the jar. - Dependent
		Roll a dice. Then roll a dice again. - Independent
12		There are a total of 14 marbles. The probability of selecting a red marble first is $\frac{10}{91}$. Then the probability of selecting a blue marble out of the 13 marbles remaining is $\frac{5}{14}$. The final probability is $\frac{5}{14}$ X $\frac{4}{13} = \frac{20}{182} = \frac{10}{91}$

Additional Information

Test Taking Tips

1) **The day before the test,** make sure you get a good night's sleep.

2) **On the day of the test,** be sure to eat a good hearty breakfast! Also, be sure to arrive at school on time.

3) **During the test:**

- **Read every question carefully.**

  - Do not spend too much time on any one question. Work steadily through all questions in the section.
  - Attempt all of the questions even if you are not sure of some answers.
  - If you run into a difficult question, eliminate as many choices as you can and then pick the best one from the remaining choices. Intelligent guessing will help you increase your score.
  - Also, mark the question so that if you have extra time, you can return to it after you reach the end of the section.
  - Some questions may refer to a graph, chart, or other kind of picture. Carefully review the graphic before answering the question.
  - Be sure to include explanations for your written responses and show all work.

- **While Answering Multiple-Choice (EBSR) questions.**

  - Select the bubble corresponding to your answer choice.
  - Read **all** of the answer choices, even if think you have found the correct answer.

- **While Answering TECR questions.**

  - Read the directions of each question. Some might ask you to drag something, others to select, and still others to highlight. Follow all instructions of the question (or questions if it is in multiple parts)

Frequently Asked Questions(FAQs)

For more information on the assessment, visit
www.lumoslearning.com/a/atlas-faqs
OR Scan the **QR Code**

Step 1 → **Visit the link given below and login to your parent/teacher account**
www.lumoslearning.com

Step 2 → **For Parent**
Click on the horizontal lines (≡) in the top right-hand corner and select **"My tedBooks"**. Place the Book Access Code and submit.

For Teacher
Click on "My Subscription" under the "My Account" menu in the left-hand side and select **"My tedBooks"**. Place the Book Access Code and submit.

Note: See the first page for access code.

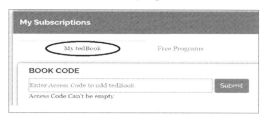

Step 3 → **Add the new book**

To add the new book for a registered student, choose the **'Student'** button and click on submit.

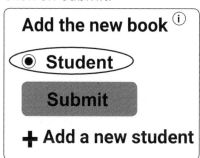

To add the new book for a new student, choose the **'Add New Student'** button and complete the student registration.

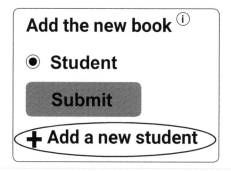

Progress Chart

Standard	Lesson	Score	Date of Completion
AAS			
7.NCC.1 7.NCC.7	Add and Subtract Rational Numbers		
7.NCC.2	Rational Numbers, Addition & Subtraction		
7.NCC.3	Additive Inverse and Distance Between Two Points on a Number Line		
7.NCC.4	Converting Between Rational Numbers and Decimals		
7.NCC.5 7.NCC.9	Solving Real World Problems		
7.NCC.6	Strategies for Adding and Subtracting Rational Numbers		
7.NCC.6	Strategies for Multiplying and Dividing Rational Numbers		
7.NCC.8	Rational Numbers as Quotients of Integers		
7.PR.1	Finding Constant of Proportionality		
7.PR.2	Unit Rates		
7.PR.3	Applying Ratios and Percents		
7.PR.4	Understanding and Representing Proportions		
7.PR.6	Represent Proportions by Equations		
7.PR.7	Significance of Points on Graphs of Proportions		
7.ALG.1	Applying Properties to Rational Expressions		
7.ALG.2	Modeling Using Equations or Inequalities		
7.ALG.3	Linear Inequality Word Problems		
7.ALG.4 7.ALG.5	Quantitative Relationships		
7.GM.1 7.GM.2	Circles		
7.GM.3	Finding Area, Volume, & Surface Area		
7.GM.4	Cross Sections of 3-D Figures		
7.GM.5	Angles		
7.GM.6	Scale Models		

Standard	Lesson	Score	Date of Completion
AAS			
7.SP.3	Mean, Median, and Mean Absolute Deviation		
7.SP.4	Mean, Median, and Mode		
7.SP.5	Sampling a Population		
7.SP.6	Describing Multiple Samples		
7.SP.7 7.SP.9	Predicting Using Probability		
7.SP.8	Understanding Probability		
7.SP.10	Using Probability Models		
7.SP.10	Probability Models from Observed Frequencies		
7.SP.10	Find the Probability of a Compound Event		

Grade **7**

Lumos Learning
Step Up Your Skills

ARKANSAS

ENGLISH
LANGUAGE ARTS LITERACY

ATLAS Practice

(((tedBook)))
ONLINE

2 Practice Tests
Personalized Study Plan

ELA Domains Literature • Informational Text • Language

Available
- At Leading book stores
- Online www.LumosLearning.com

Notes

Test Taking Tips

1) **The day before the test,** make sure you get a good night's sleep.

2) **On the day of the test,** be sure to eat a good hearty breakfast! Also, be sure to arrive at school on time.

3) **During the test:**

- **Read every question carefully.**

  - Do not spend too much time on any one question. Work steadily through all questions in the section.
  - Attempt all of the questions even if you are not sure of some answers.
  - If you run into a difficult question, eliminate as many choices as you can and then pick the best one from the remaining choices. Intelligent guessing will help you increase your score.
  - Also, mark the question so that if you have extra time, you can return to it after you reach the end of the section.
  - Some questions may refer to a graph, chart, or other kind of picture. Carefully review the graphic before answering the question.
  - Be sure to include explanations for your written responses and show all work.

- **While Answering Multiple-Choice (EBSR) questions.**

  - Select the bubble corresponding to your answer choice.
  - Read **all** of the answer choices, even if think you have found the correct answer.

- **While Answering TECR questions.**

  - Read the directions of each question. Some might ask you to drag something, others to select, and still others to highlight. Follow all instructions of the question (or questions if it is in multiple parts)

Frequently Asked Questions(FAQs)

For more information on the assessment, visit
www.lumoslearning.com/a/atlas-faqs
OR Scan the **QR Code**

Step 1 → **Visit the link given below and login to your parent/teacher account**
www.lumoslearning.com

Step 2 → <u>For Parent</u>
Click on the horizontal lines (≡) in the top right-hand corner and select **"My tedBooks"**. Place the Book Access Code and submit.

<u>For Teacher</u>
Click on "My Subscription" under the "My Account" menu in the left-hand side and select **"My tedBooks"**. Place the Book Access Code and submit.

Note: See the first page for access code.

Step 3 → **Add the new book**

To add the new book for a registered student, choose the **'Student'** button and click on submit.

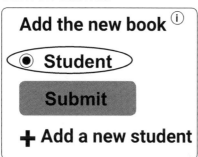

To add the new book for a new student, choose the **'Add New Student'** button and complete the student registration.

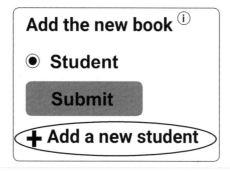

Progress Chart

Standard	Lesson	Score	Date of Completion
AAS			
7.NCC.1 7.NCC.7	Add and Subtract Rational Numbers		
7.NCC.2	Rational Numbers, Addition & Subtraction		
7.NCC.3	Additive Inverse and Distance Between Two Points on a Number Line		
7.NCC.4	Converting Between Rational Numbers and Decimals		
7.NCC.5 7.NCC.9	Solving Real World Problems		
7.NCC.6	Strategies for Adding and Subtracting Rational Numbers		
7.NCC.6	Strategies for Multiplying and Dividing Rational Numbers		
7.NCC.8	Rational Numbers as Quotients of Integers		
7.PR.1	Finding Constant of Proportionality		
7.PR.2	Unit Rates		
7.PR.3	Applying Ratios and Percents		
7.PR.4	Understanding and Representing Proportions		
7.PR.6	Represent Proportions by Equations		
7.PR.7	Significance of Points on Graphs of Proportions		
7.ALG.1	Applying Properties to Rational Expressions		
7.ALG.2	Modeling Using Equations or Inequalities		
7.ALG.3	Linear Inequality Word Problems		
7.ALG.4 7.ALG.5	Quantitative Relationships		
7.GM.1 7.GM.2	Circles		
7.GM.3	Finding Area, Volume, & Surface Area		
7.GM.4	Cross Sections of 3-D Figures		
7.GM.5	Angles		
7.GM.6	Scale Models		

Standard	Lesson	Score	Date of Completion
AAS			
7.SP.3	Mean, Median, and Mean Absolute Deviation		
7.SP.4	Mean, Median, and Mode		
7.SP.5	Sampling a Population		
7.SP.6	Describing Multiple Samples		
7.SP.7 7.SP.9	Predicting Using Probability		
7.SP.8	Understanding Probability		
7.SP.10	Using Probability Models		
7.SP.10	Probability Models from Observed Frequencies		
7.SP.10	Find the Probability of a Compound Event		

Grade 7

Lumos Learning
Step Up Your Skills

ARKANSAS

ENGLISH
LANGUAGE ARTS LITERACY

ATLAS Practice

(((tedBook)))
ONLINE

2 Practice Tests
Personalized Study Plan

ELA Domains | Literature • Informational Text • Language

Available
- At Leading book stores
- Online www.LumosLearning.com

Made in the USA
Columbia, SC
29 February 2024